HORIZONTAL-BRANCH AND UV-BRIGHT STARS

The Proceedings of a Meeting Held at
NASA, Goddard Spaceflight Center
June 9-10, 1985

EDITED BY

A. G. Davis Philip

Van Vleck Observatory and Union College

Van Vleck Observatory Contribution No. 4

L. Davis Press, Inc.
Schenectady, N. Y.
1985

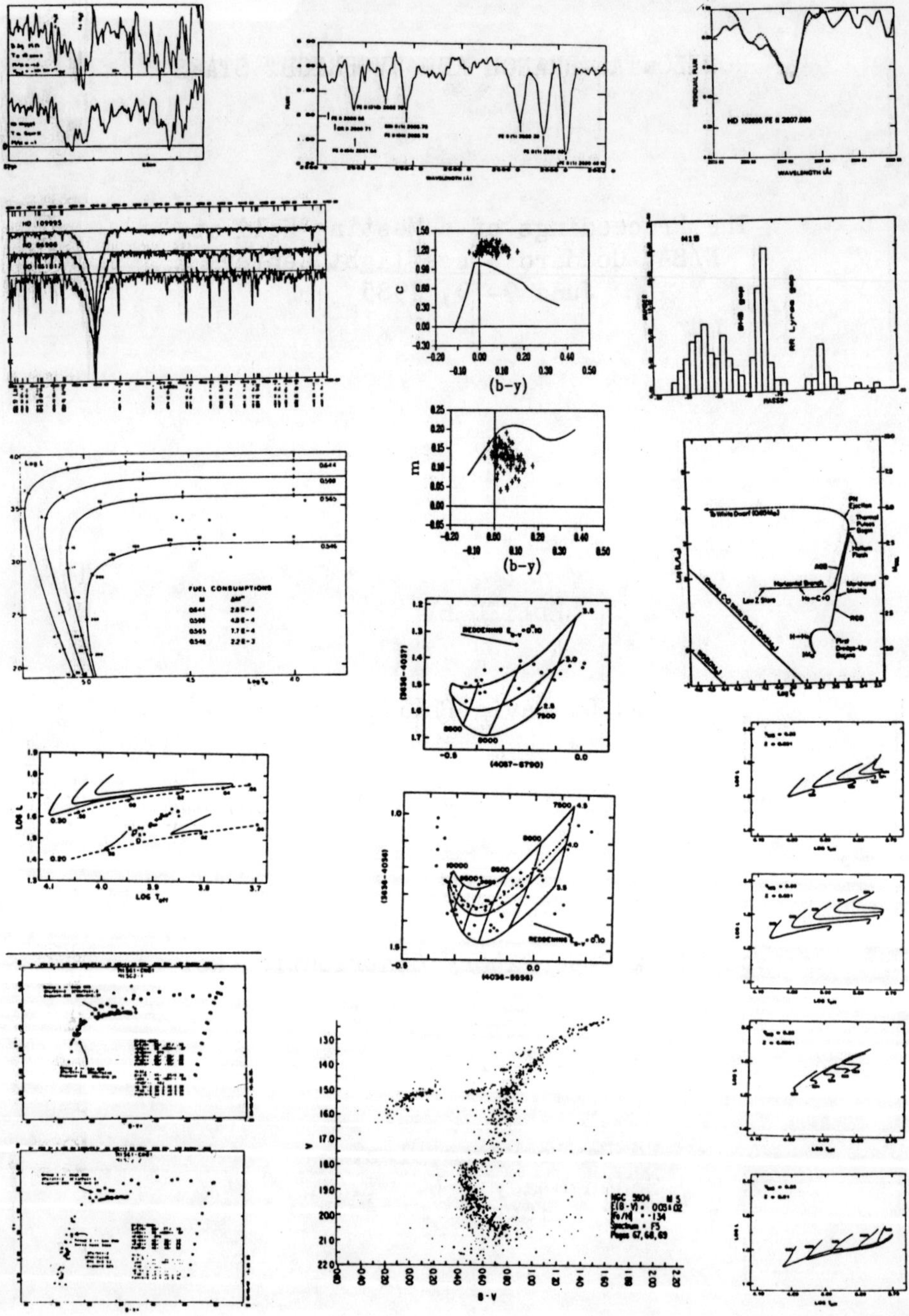
HD 109995
(b-y)
C
m1
Log L
FUEL CONSUMPTIONS
0.644 2.0E-4
0.598 4.8E-4
0.565 7.7E-4
0.546 2.2E-3
REDDENING
(4037-6790)
(4036-5656)
LOG L
LOG T
B-V
V
NGC 5904 M 5

The Participants at the NASA Meeting

From left to right: David Leckrone, Saul Adelman, Ken Janes, Davis Philip, Alvio Renzini and Allen Sweigart.

TABLE OF CONTENTS

PREFACE

A meeting on the subject of Horizontal-Branch and UV-Bright Stars was held at the Goddard Spaceflight Center, NASA on June 9 - 10, 1985. This was the fourth meeting of a consortium, the members of which are interested in the problems of evolved Population II blue stars. They are:

Saul J. Adelman	*	NASA/Goddard and The Citadel	High Dispersion Spectroscopy and Spectrophotometry
Klaas S. de Boer		Kapteyn Astronom. Institute	Ultraviolet Spectroscopy and Photometry
Pierre Demarque		Yale University Observatory	Theory of Stellar Evolution
Donald S. Hayes		Kitt Peak National Obs.	Photometry and Spectrophotometry
Kenneth Janes	*	Boston University	Photometry of Pop II Stars
Keiichi Kodaira		Tokyo Observatory	High Dispersion Spectroscopy
David S. Leckrone	*	NASA/Goddard	High Dispersion UV Spectroscopy and Photometry
Ruth C. Peterson		Whipple Observatory Smithsonian Inst.	High Dispersion Spectroscopy and Stellar Rotation
A. G. Davis Philip	*	Van Vleck Obs. and Union College	Photometry and Spectrophotometry
Alvio Renzini	*	Bologna Observatory	Theory of Stellar Evolution
Allen V. Sweigart	*	NASA/Goddard	Theory of Stellar Evolution

The persons whose names are marked with an asterisk attended this NASA meeting. Each member of the consortium has contributed a paper to the proceedings. In addition there are four papers authored or co-authored by persons who are not members of the consortium. The discussion following the oral papers was recorded, transcribed and then mailed to those present at the meeting for correction. Individual authors have been made responsible for obtaining permission to copy material in other publications. The manuscripts have been retyped in order to make all the papers uniform in style. The order of the spoken papers is the order in which they were presented at the meeting. The contributed papers have been added at the appropriate place according to the subject of each paper. The figure on the cover was taken from the paper "The Identification and Study of Horizontal-Branch Stars" by Philip.

The Laboratory for Astronomy and Solar Physics, NASA Goddard Spaceflight Center provided space for the two day June meeting and we thank Dave Leckrone and his associates for arranging matters there. The index was prepared on an Apple IIc computer (courtesy of K. W. Philip). The main part of the preparation of the manuscripts was done on an IBM PC.

A. G. Davis Philip

Editor

August, 1985
Schenectady, New York

THE PROCEEDINGS

THE EVOLUTION OF HORIZONTAL-BRANCH STARS

Allen V. Sweigart

Laboratory for Astronomy and Solar Physics
NASA Goddard Space Flight Center

ABSTRACT. The role of the horizontal branch in the study of globular-cluster compositions and ages is discussed. A new grid of horizontal-branch evolutionary sequences is presented for helium abundances between 0.20 and 0.30 and heavy-element abundances between 0.0001 and 0.01. These sequences include the envelope helium enrichment due to convective dredge-up during the earlier subgiant-branch phase and use the proper core mass for each composition. Methods for determining the globular-cluster helium abundance from the properties of horizontal-branch stars are briefly reviewed. Finally the implications of the Sandage period-shift effect for both globular-cluster compositions and horizontal-branch models are examined on the basis of some new stellar-evolution calculations.

1. INTRODUCTION

The study of horizontal-branch (HB) stars in galactic globular clusters is important for a number of astrophysical problems. One example is the problem of determining the chemical composition of the globular clusters. The HB is useful for this problem because both the morphology of the theoretical HB sequences in the HR diagram and the evolutionary lifetimes are sensitive to the chemical composition. In addition, since the HB stars are often among the hotter members of the globular clusters, they will be prime candidates for spectroscopic abundance determinations from ultraviolet observations with the Space Telescope (ST). The abundances derived for HB stars reflect not only the primordial compositions of the globular clusters but also the effects of mixing during the preceding red-giant phase. Observations clearly indicate that nucleosynthesized material is mixed into the envelopes of many red-giant stars, thereby leading, for example, to substantial changes in the individual CNO abundances and isotope ratios. Reliable HB abundances would be quite helpful for better understanding the characteristics and extent of this mixing process.

A. G. Davis Philip, (ed.)
Horizontal-Branch and UV-Bright Stars 3-18

Another example of the HB's importance concerns the use of the HB in dating the globular clusters. At the present time the best method for determining the globular-cluster ages is based on the observed luminosity difference ΔM_{TO}^{RR}(bol) between the HB and the main-sequence turnoff. This luminosity difference can be calibrated as a function of the age from theoretical evolutionary sequences. Unfortunately this calibration is also a function of the helium abundance Y and the heavy-element abundance Z. Using the calibration given in equation (9) of Iben and Renzini (1984), one finds that an uncertainty in Y of ± 0.02 or in log Z (or [Fe/H]) of ± 0.5 leads to an uncertainty of 10 percent in the age. Thus the problem of the globular-cluster ages is directly tied to the problem of the globular-cluster abundances. A more serious concern with the ΔM_{TO}^{RR}(bol) dating method arises from the strong sensitivity of the age to the theoretical HB luminosity. An error of only 0.1 mag in the luminosity of the HB models causes an error of 1.5×10^9 years in the age. Thus the derived globular cluster ages are strongly dependent on the accuracy of the stellar-evolution calculations for the HB phase.

The HB is also the locus of the RR Lyrae variables. This fact is of considerable importance, first, because the RR Lyrae variables provide a wealth of additional observational data and, second, because one can apply the results of pulsation theory. Combining stellar-evolution theory and pulsation theory with the observations can lead to stringent tests of the available HB models. For example, recent observations by Sandage, Katem and Sandage (1981) and Sandage (1981, 1982) have shown that there is a shift in the pulsation periods of RR Lyrae variables at a given effective temperature between globular clusters of different metallicity. These observations have raised significant questions about how well the current theoretical models reproduce the properties of actual HB stars.

In summary, the study of the HB is fundamentally an attempt to understand the early chemical and dynamical evolution of the galaxy through determining the globular-cluster abundances and ages as well as a means to test the results of stellar-evolution theory and therefore our understanding of the HR diagram morphology of the globular clusters.

The main purposes of this paper are to present a new grid of theoretical HB sequences for different helium and heavy-element abundances (Section 3) and to discuss the use of such sequences in determining the globular-cluster helium abundance (Section 4). In Section 5 we address the presently unresolved problem posed by the Sandage period-shift observations in more detail. Before presenting the new HB sequences, however, we briefly discuss in the following section how the globular-cluster age derived from the main-sequence turnoff luminosity depends on the globular-cluster composition.

2. GLOBULAR-CLUSTER AGES

The first key project recommended by the ST Working Group on Stars and Star Clusters is to determine the ages of the globular clusters to within 10 percent from observations of the main-sequence turnoff luminosity L_{TO}. ST can be used to obtain the turnoff luminosity in several ways, e.g., 1) by observing the white dwarfs in the globular clusters and using their cooling curves as standard candles for determining the distance moduli and 2) by fitting the globular-cluster main sequences to the subdwarf main sequence as defined by ST parallax observations of nearby subdwarfs. However, even if these methods yielded a precise value for L_{TO}, the ages derived for the globular clusters would still depend on the accuracy of the current stellar models and on the globular-cluster composition. Let us set aside any questions about the stellar models and instead focus on the composition dependence of the globular-cluster ages at a fixed turnoff luminosity. This dependence is illustrated in Figures 1 and 2, where we plot log L_{TO} in solar units as a function of the logarithm of the globular-cluster age in units of 10^9 years. The data in these figures were taken from the isochrones of Ciardullo and Demarque (1977), which, in turn, are based on the main-sequence turnoff tracks of Mengel et al. (1979).

Figure 1 shows the dependence of log L_{TO} on log age for different values of the helium abundance at a metallicity of Z = 0.001. It is immediately apparent from this figure that the age corresponding to a given turnoff luminosity is sensitive to the helium abundance. The uncertainty Δt in the age at fixed L_{TO} resulting from an uncertainty ΔY in the helium abundance is given by

$$\Delta t/t = -1.3\ \Delta Y. \qquad (1)$$

Hence, if we hope to tie down the globular-cluster ages to within 10 percent, we must know the helium abundance to within ± 0.08. At first glance this requirement may not seem particularly severe. However, it should be remembered that there will also be uncertainties in both the turnoff luminosity and the metallicity. If the total uncertainty in the age is to be kept within 10 percent, then the constraint on the maximum permitted uncertainty in the helium abundance is substantially more stringent.

The dependence of log L_{TO} on log age for different metallicities is shown in Figure 2 for a helium abundance of Y = 0.30. We find that the age derived from a given turnoff luminosity is also quite sensitive to the assumed globular-cluster metallicity. In this case the uncertainty Δt in the age at fixed L_{TO} is related to the uncertainty $\Delta \log Z$ in the metallicity by

$$\Delta t/t = -0.4\ \Delta \log Z, \qquad (2)$$

which implies that the uncertainty in log Z and thus in [Fe/H] must be less than ±0.2 if the globular-cluster ages are to be known to within 10 percent. As before, this maximum uncertainty in the metallicity assumes

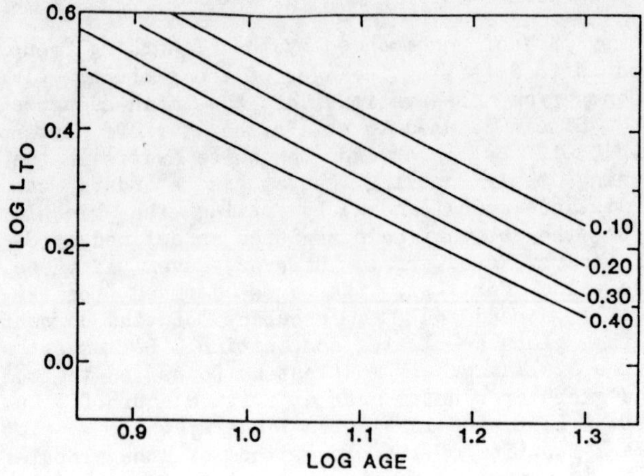

Fig. 1. Dependence of the main-sequence turnoff luminosity L_{TO} in solar units on the globular-cluster age in units of 10^9 years at a metallicity of Z = 0.001. Each line is labelled by its value of the main-sequence helium abundance.

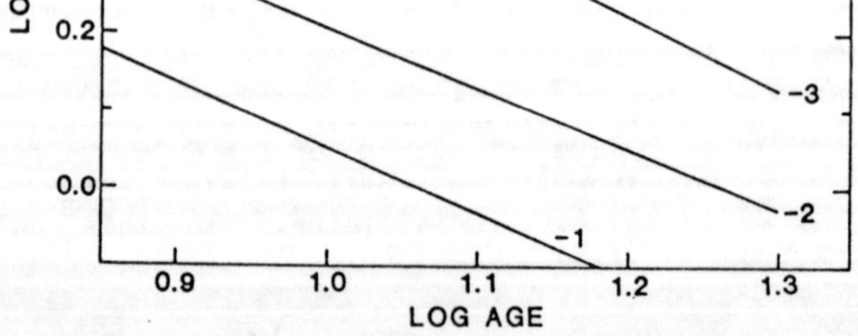

Fig. 2. Dependence of the main-sequence turnoff luminosity L_{TO} in solar units on the globular-cluster age in units of 10^9 years for a main-sequence helium abundance of 0.30. Each line is labelled by its value of log Z.

that all other uncertainties are negligible. Since this will clearly not be the case, it is essential to obtain as accurate values as possible for the heavy-element abundances, particularly CNO, if the goal of the above ST key project is to be realized.

3. HORIZONTAL-BRANCH SEQUENCES

As indicated in Section 1, studies of the HB in globular clusters depend on the availability of suitable stellar models. For this reason we have constructed an improved grid of 45 evolutionary sequences including semiconvection for HB stars having effective temperatures in the range from roughly log T_{eff} = 3.7 to 4.0. Each sequence was evolved from the zero-age horizontal branch (ZAHB) to the point where the central helium abundance was depleted to 0.05. Thus the final phase of core-helium burning was not included. This final phase is complicated theoretically by the occurrence of a composition instability which causes the convective core to grow substantially in mass (Sweigart and Demarque 1973, Castellani et al. 1985).

In this section we wish to illustrate how these new evolutionary sequences depend on both the helium and heavy-element abundances. Before doing this, however, we wish to comment on the principal ways in which the present sequences differ from most previous sequences. These differences include:

a) The enrichment in the envelope helium abundance due to convective dredge-up along the subgiant branch has been taken into account. During the subgiant-branch phase the convective envelope penetrates inward into regions where the hydrogen abundance was partially depleted during the preceding main-sequence phase (Faulkner and Iben 1967). The resulting convective dredge-up enriches the envelope helium abundance by an amount ΔY_s that ranges typically from 0.01 to 0.02 (Sweigart and Gross 1978). Consequently we will distinguish in the following discussion between the primordial main-sequence helium abundance Y_{MS} and the envelope helium abundance Y_{HB} (= $Y_{MS} + \Delta Y_s$) in the HB models.

b) The core mass M_c appropriate for the composition has been used in each sequence. The value of M_c for each case was determined as follows. For each composition we first determined the main-sequence turnoff mass M_{TO} for an age of 15 x 10^9 years. This choice for the age represents a reasonable estimate for the globular-cluster ages, but the exact value is not crucial, since M_c is only mildly dependent on mass. Given M_{TO}, we then obtained the value of M_c by interpolating in the red-giant evolutionary sequences of Sweigart and Gross (1978). In most previous sequences M_c was regarded as a free parameter, and thus these sequences, unlike the present ones, are usually not self-consistent.

c) Evolutionary sequences have been computed for the metallicity Z = 0.01. Thus the present results extend the available theoretical sequences into the domain of the metal-rich RR Lyrae variables.

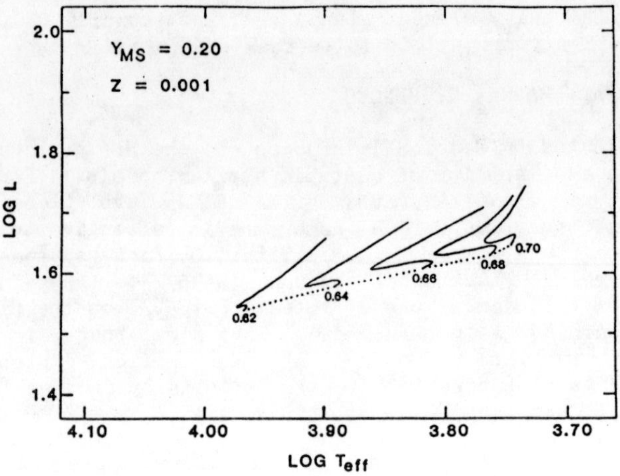

Fig. 3. HB evolutionary sequences for Y_{MS} = 0.20 and Z = .001. Each track is labelled by its mass in solar units. The ZAHB locus is indicated by the dotted curve. The envelope helium abundance and the core mass for these sequences are Y_{HB} = 0.2170 and M_c = 0.4954 $M_\odot$.

Sequences for five values of the mass have been computed for each combination of the following values of Y_{MS} and Z:

$$Y_{MS} = 0.20,\ 0.25,\ 0.30$$

and

$$Z = 0.0001,\ 0.001,\ 0.01.$$

The dependence of these sequences on the helium abundance is illustrated in Figures 3, 4 and 5, where we plot the sequences with Y_{MS} = 0.20, 0.25 and 0.30, respectively, for the metallicity Z = 0.001. Comparing these figures shows that an increase in Y_{MS} leads to 1) higher HB luminosities, 2) longer blueward loops and 3) larger deviations in luminosity from the ZAHB curves. These three effects can be understood readily. Increasing Y_{MS} leads to a larger mean molecular weight within the envelope and thus to a reduction in the envelope gas pressure. In order to maintain hydrostatic equilibrium, the envelope must contract until the temperature rises sufficiently for the outward gas pressure to balance once again the inward weight of the overlying layers. This temperature rise increases the energy output of the hydrogen shell and hence the HB luminosity. As a result, the hydrogen shell then advances outward more rapidly in mass, thereby increasing the amount by which the

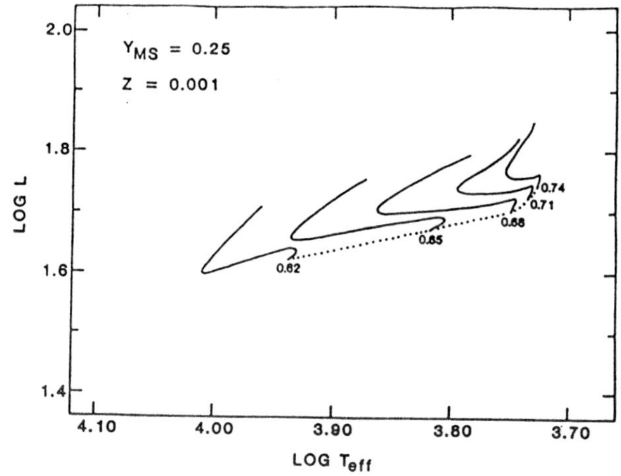

Fig. 4. HB evolutionary sequences for Y_{MS} = 0.25 and Z = 0.001. Each track is labelled by its mass in solar units. The envelope helium abundance and core mass for these sequences are Y_{HB} = 0.2636 and M_c = 0.4858 $M_{\odot}$.

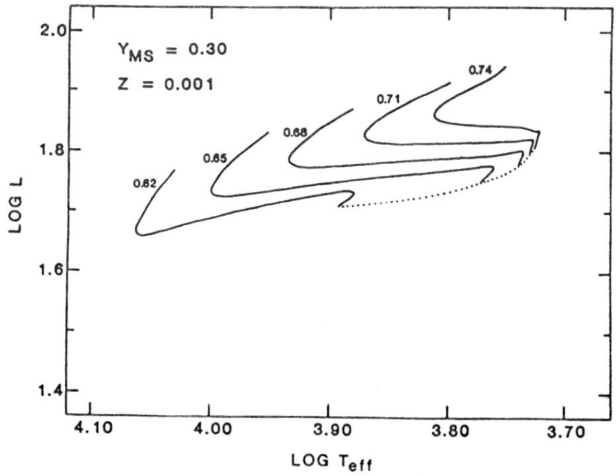

Fig. 5. HB evolutionary sequences for Y_{MS} = 0.30 and Z = 0.001. Each track is labelled by its mass in solar units. The envelope helium abundance and core mass for these sequences are Y_{HB} = 0.3101 and M_c = 0.4761 $M_{\odot}$.

helium-core mass grows during the HB phase. It is well-known that HB models with larger core masses have higher effective temperatures and higher luminosities (see, e.g., Sweigart and Gross 1976). Thus the second and third effects noted above are a direct consequence of the greater growth in the helium-core mass found at the larger helium abundances.

In order to illustrate the metallicity dependence of the present sequences, we have plotted the sequences for Z = 0.0001 and 0.01 in Figures 6 and 7, respectively. These figures, together with Figure 4, all refer to a helium abundance of $Y_{MS} = 0.25$, and from them we conclude that an increase in Z leads to lower HB luminosities and longer blueward loops. Again these two effects can be understood readily.

HB models with higher metallicities have smaller core masses at the ZAHB phase (Rood 1972, Sweigart and Gross 1978) and larger envelope radiative opacities. A smaller core mass lowers both the hydrogen- and helium-burning rates and hence the luminosity in HB models with the same effective temperature (see, e.g., Sweigart and Gross 1976). A larger envelope opacity forces an HB model to shift redward in the HR diagram. The only way to move the model back to the original effective temperature (without changing the composition or core mass) is to reduce the envelope mass. Thus, if one examines HB models of different metallicity but the same effective temperature, one expects the higher metallicity models to have smaller masses. This expectation is confirmed by comparing the tracks in Figure 7 with those in Figures 4 or 6. A smaller mass decreases the weight of the envelope and thus the pressure at the hydrogen shell which, in turn, lowers the shell temperature and consequently the hydrogen-burning rate. Thus both the smaller core mass and the larger envelope opacity in the higher metallicity models lead to a reduction in the HB luminosity.

The extent of the blueward loop depends on the amount by which the envelope mass decreases during the HB phase. As just discussed, the higher metallicity models have substantially smaller envelope masses at a given effective temperature. Therefore a given decrease in the envelope mass of these models represents a considerably greater perturbation to the envelope's structure. This fact explains why the higher metallicity tracks, such as those in Figure 7, have longer blueward loops even though their luminosities are somewhat smaller.

4. GLOBULAR-CLUSTER HELIUM ABUNDANCE

The properties of HB sequences such as those in Section 3 can be used in several ways to estimate the globular-cluster helium abundance. From pulsation theory one can determine the pulsational characteristics of the HB models occupying the instability strip. Certain characteristics, such as the width of the instability strip and the location of the blue edge, can then be used as helium determinants (see, e.g., Iben and Renzini 1984). Another method is based on the ratio R,

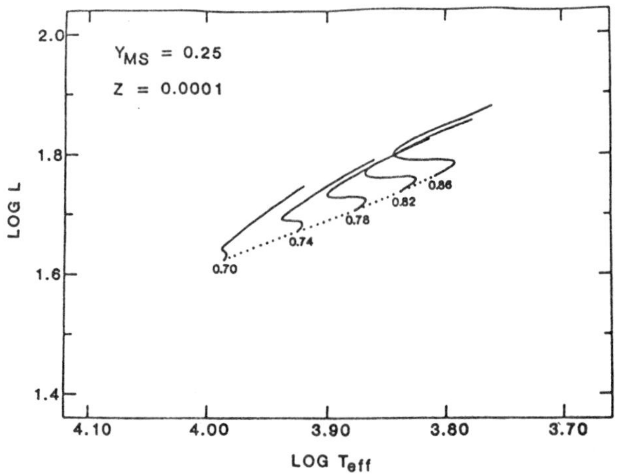

Fig. 6. HB evolutionary sequences for Y_{MS} = 0.25 and Z = 0.001. Each track is labelled by its mass in solar units. The envelope helium abundance and core mass for these sequences are Y_{HB} = 0.2588 and M_c = 0.4978 $M_\odot$.

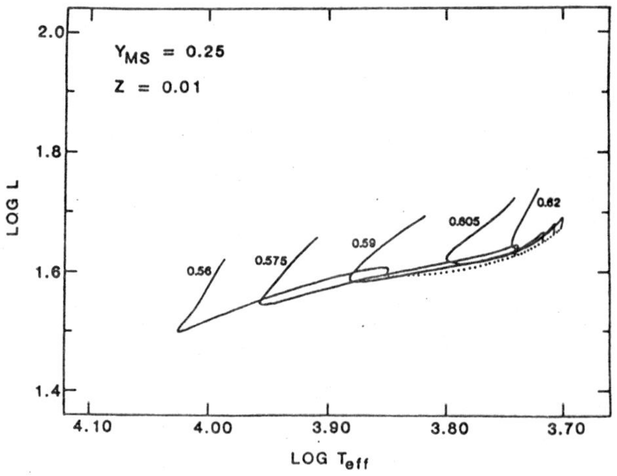

Fig. 7. HB evolutionary sequences for Y_{HB} = 0.25 and Z = 0.01. Each track is labelled by its mass in solar units. The envelope helium abundance and core mass for these sequences are Y_{HB} = 0.2698 and M_c = 0.4772 $M_\odot$.

defined by

$$R = t_{HB}/t_{RG}, \qquad (3)$$

between the HB lifetime t_{HB} and the red-giant lifetime t_{RG} above the luminosity level of the HB (Iben 1968). Observationally this ratio is equivalent to the number ratio between HB and red-giant stars. The R-method has recently been carefully applied by Buzzoni et al. (1983) to 15 globular clusters. The helium abundance derived for these clusters was 0.23 ± 0.02 with no indication of a dependence on metallicity. In using the R-method one should be aware of the possible observational and theoretical uncertainties. Observationally the helium abundance given by the R-method could be affected by incompleteness in the number counts as can happen if the bluest HB stars are missed or by difficulties in separating the asymptotic-giant-branch from the red-giant-branch stars. Theoretically there could be possible uncertainties associated with the current values of the core mass (and its dependence on rotation), the extent of semiconvection and the length of time spent in the final central helium-burning phase.

Another property of the HB that is sensitive to the helium abundance is the luminosity difference between the bluest and reddest HB stars. The variation of the ZAHB luminosity with effective temperature is shown in Figure 8 for helium abundances between Y_{MS} = 0.10 and 0.40 at a metallicity of Z = 0.001. The data in Figure 8 was obtained from the HB models of Sweigart and Gross (1976). For each composition the proper value of the core mass was used, as interpolated from Sweigart and Gross (1978). We see from Figure 8 that the drop-off in the luminosity in going blueward along the ZAHB is considerably greater at higher helium abundances. The reasons for this drop-off can be understood if we realize that the hydrogen-burning shell becomes progressively weaker in going blueward along the ZAHB. In the redder ZAHB models, say, those near the instability strip, the hydrogen-burning shell makes a major contribution to the surface luminosity, and we find that the luminosity increases with helium abundance for the reasons outlined in the previous section. In the bluest ZAHB models, say, those with log T_{eff} > 4.2, the hydrogen-shell source is rapidly weakening, and thus the surface luminosity is then dominated by the helium burning in the core. At sufficiently high effective temperatures the hydrogen shell is effectively extinguished. In this case the luminosity decreases with increasing helium abundance, since the higher helium abundance models have smaller core masses and therefore lower helium-burning rates. This is why the Y_{MS} = 0.10 curve intersects the 0.20 and 0.30 curves at log T_{eff} between 4.2 and 4.3 in Figure 8. In using Figure 8 as a helium determinant one clearly needs to determine accurate bolometric luminosities for HB stars with effective temperatures that equal or exceed 20,000 °K. At lower effective temperatures, i.e., at log T_{eff} < ∽ 4.0, it would be very difficult to distinguish between the variation of the ZAHB luminosity with effective temperature at different helium abundances. In actually applying Figure 8 one would need to include the effects of evolution away from the ZAHB. Evolution would modify the curves in Figure 8 somewhat but would not basically change the above discussion. Alternatively the curves in

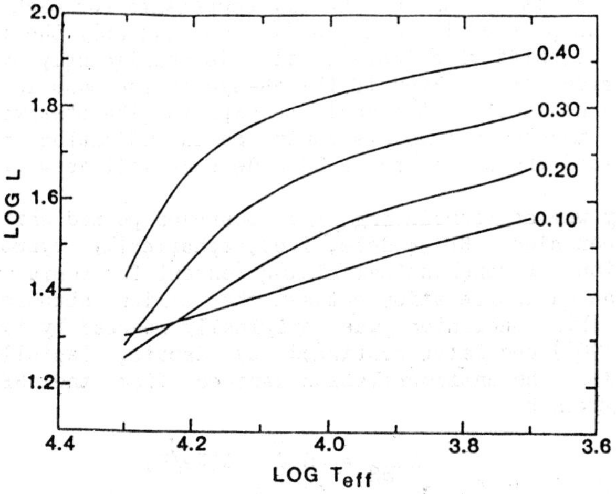

Fig. 8. Dependence of the ZAHB luminosity in solar units on the effective temperature for different helium abundances at a metallicity of Z = 0.001. Each curve is labelled by its value of Y_{MS}.

Figure 8 could be compared with the lower luminosity envelope of observed HB's, since this envelope should correspond to the ZAHB.

5. SANDAGE PERIOD-SHIFT EFFECT

The final topic to be discussed in this paper is the Sandage period-shift effect mentioned in Section 1. This effect is important because it provides a basic test of current HB models and because it could have significant implications for the study of globular-cluster abundances. Essentially the period-shift effect refers to the fact that the RR Lyrae variables in the metal-poor globular clusters are observed to have longer pulsation periods than those at the same effective temperature in the metal-rich clusters.

The observed period shift between the metal-poor and metal-rich clusters can be understood if the metal-poor RR Lyrae variables are more luminous than the metal-rich ones. If there is no change in the RR Lyrae mass between the metal-poor and metal-rich clusters, then the observed period shift implies an increase in the HB luminosity of 0.35 mag. per one dex decrease in [Fe/H]. As pointed out in Section 1, such a variation in the HB luminosity would substantially alter the globular-cluster ages derived from the ΔM_{TO}^{RR}(bol) dating method. If the change in the RR Lyrae mass is taken into account, the required luminosity increase would be even greater.

Current HB models such as those in Figures 4, 6 and 7 predict an increase in both the RR Lyrae luminosity and mass at a given effective temperature with decreasing metallicity. Unfortunately these changes in the RR Lyrae luminosity and mass nearly compensate for each other as far as the pulsation period is concerned. Consequently the theoretically predicted period shift with metallicity is vanishingly small compared with the observed shift. Even if the change in the mass is ignored, the luminosity change is still too small to explain the observations. This failure of the theoretical models could be an indication that we do not really understand the structure of HB models as well as we think.

The only way of reproducing the observed period shift within the context of canonical HB models, i.e., spherically symmetric models without rotation or nonclassical mixing (except for semiconvection), is to postulate an anticorrelation between the helium abundance and the metallicity. This conclusion was originally reached by Sandage, Katem and Sandage (1981) and later confirmed by Caputo, Castellani and di Gregorio (1983). The anticorrelation derived from the HB models in Section 3 is given by

$$\Delta Y_{MS} = -0.09 \ \Delta[Fe/H] \qquad (4)$$

(Sweigart, Renzini and Tornambe 1985). Such an anticorrelation seems very difficult to accept, since it would require a very implausible protogalactic nucleosynthetic history. Moreover, the helium abundances derived by Buzzoni et al. (1983) from the R-method give no hint of such an anticorrelation.

A number of possible causes for the observed period shift have been investigated recently by Sweigart, Renzini and Tornambe (1985). We have studied, for example, whether possible uncertainties in the red-giant input physics that determines the metallicity dependence of the core mass might be responsible for the discrepancy between the observed and predicted period shifts. These red-giant calculations indicate, however, that even large changes in the input physics are unlikely to resolve this discrepancy. One effect not fully included in previous studies of the period shift is the effect of evolution away from the ZAHB. Our results show that including the HB evolution, as given by the sequences in Section 3, does increase the predicted period shift somewhat but not enough to explain the observations. One effect which would lead to a larger period shift is an increase in the metal opacity at temperatures around a few million degrees. Test models for the ZAHB phase suggest that an increase by a factor of 5 in this metal opacity would produce the observed period shift. Whether such a large increase in current metal opacities is possible is uncertain, and thus it is not clear if this explanation for the period shift is tenable. At the present time there is no fully satisfactory explanation of the period-shift discrepancy. For this reason the HB must continue to be the focus of active research interest.

REFERENCES

Buzzoni, A., Fusi Pecci, F., Buonanno, R. and Corsi, C. E. 1983 Astron. Astrophys. 128, 94.
Caputo, F., Castellani, V. and di Gregorio, R. 1983 Astron. Astrophys. 123, 141.
Castellani, V., Chieffi, A., Pulone, L. and Tornambe, A. 1985 Astrophys. J., in press.
Ciardullo, R. B. and Demarque, P. 1977 Trans. Yale Univ. Obs. vol. 33.
Faulkner, J. and Iben, I., Jr. 1967 Nature 215, 44.
Iben, I., Jr. 1968 Nature, 220, 143.
Iben, I., Jr. and Renzini, A. 1984 Physics Reports 105, 329.
Mengel, J. G., Sweigart, A. V., Demarque, P. and Gross, P. G. 1979 Astrophys. J. Suppl. 40, 733.
Rood, R. T. 1972 Astrophys. J. 177, 681.
Sandage, A. 1981 Astrophys. J. 248, 161.
Sandage, A. 1982 Astrophys. J. 252, 553.
Sandage, A., Katem, B. and Sandage, M. 1981 Astrophys. J. Suppl. 46, 41.
Sweigart, A. V. and Demarque, P. 1973 in IAU Coll. No. 21, Variable Stars in Globular Clusters and Related Systems, J. D. Fernie, ed., Reidel, Dordrecht, p. 221.
Sweigart, A. V. and Gross, P. G. 1976 Astrophys. J. Suppl. 32, 367.
Sweigart, A. V. and Gross, P. G. 1978 Astrophys. J. Suppl. 36, 405.
Sweigart, A. V., Renzini, A. and Tornambe, A. 1985 Astrophys. J., to be submitted.

DISCUSSION

PHILIP: Allen, as I remember, the stars spend more time on the lower part of the track than on the upper, right?

SWEIGART: Yes. The evolution is slower on the lower part of the track, especially near the blue end of the blueward loop.

JANES: What about your most metal-poor track? Presumably it gradually increases along the track.

SWEIGART: Yes, the evolution gradually speeds up as the star moves redward to the AGB.

LECKRONE: I am trying to relate the tracks to what you would observe in order to answer what question. Does one get helium abundance and metallicity independently and then use these estimates to constrain your models?

SWEIGART: Let me answer one aspect of your question, by making some comments on the helium abundance. One way in which these models can be used to determine the helium abundance is based on a comparison between the theoretical HB and red-giant-branch lifetimes. The ratio, R, of these lifetimes is primarily a function of the helium abundance. If one uses the observed number counts to estimate the value of R, the derived helium abundance turns out to be about 23 percent.

JANES: Presumably you have lifetime information from the tracks.

SWEIGART: Yes, that is right.

LECKRONE: How did Sandage determine Fe abundances? What is the source of the delta P, delta Mv relation?

SWEIGART: Sandage based his Fe abundance on the metallicity scale of Zinn (1960, Astrophys. J. Suppl. 42, 19). There are two other derivations of this relation. One is by Roger Bell (1984, private communication), who found $\Delta \log P = -0.085\ \Delta[Fe/H]$. The other is by Bingham et al. (1984, Monthly Notices Roy. Astron. Soc. 209, 765), who found $\Delta \log P = -0.094\ \Delta[Fe/H]$. So there is some variation in the numerical coefficient in this relation. Nevertheless a value of ~ 0.1 for this coefficient seems well-established and is clearly in disagreement with the theory.

JANES: The CNO abundance may have some bearing on the Fe/H abundance.

SWEIGART: Suppose you take an HB model for a given metallicity with [CNO/Fe] = 0. Suppose that you now increase the CNO by a factor of ten while leaving the Fe abundance unchanged. What happens? First, the efficiency of the hydrogen burning will go up by a factor of ten, since the hydrogen shell in HB models operates mainly by the CNO cycle. Second, the opacity at temperatures around a few million degrees will increase. At lower or higher temperatures the opacity is not very sensitive to the metal abundance. Let us now take another model in which both the CNO and Fe abundances are increased by the same factor of ten. What will be the difference between the model with just CNO enhanced and the model with both CNO and Fe enhanced? Well, the efficiency of the CNO cycle will be identical in both models. Since the CNO elements are the main contributors to the metal opacity around a few million degrees, the increase in the opacity will also be nearly identical in both models. So the structure and observational characteristics of a model with just CNO enhanced are going to be the same in first approximation as those of a model with both CNO and Fe enhanced. Therefore, if the theoretical models do not give the observed period shift when both CNO and Fe are varied together, one would not expect an increase in CNO by itself to give a different result.

JANES: I am trying to envision some combination of doing things out of step with Fe/H.

ADELMAN: In the atmospheres, the iron abundance is far more important than that of CNO.

JANES: I am trying to think of how to solve the period shift problem. If at one CNO you have a CNO/Fe ratio and at another metallicity you have a different CNO/Fe ratio, what do you juggle to make it do something. But it goes in the wrong direction, I think.

SWEIGART: There is a certain predicted period shift associated with a given change in the metallicity. The only way of increasing that shift is by increasing the change in metallicity. If you do not increase the change in Fe abundance, then you have to increase the change in the CNO abundance. What we then want to have is [CNO/Fe] = 0 in the very metal-poor models and [CNO/Fe] > 0 in the metal-rich models. However, the observational data on Population II CNO abundances suggest just the opposite.

RENZINI: The possibility might be that there is no primordial variation in helium, but that helium anticorrelates with metals in HB stars because of a stronger helium dredge-up at lower metallicities. However models show the opposite trend, as Sweigart has discussed a moment ago, and the amount of helium dredge up is very small anyway. So, it seems to me unlikely that helium variations are the cause of the period shift. Then we tried to see whether changes in the metal opacity could produce the period shift. At around one million degrees the metal contribution to opacity is very important, but you have many elements, many ionization stages, many excitations and lots of transitions and one must take them all into account. The business of computing opacities is then most complicated around one million degrees. What we found was that one should increase the metal opacity by at least a factor of five, relative to the Los Alamos opacities to get the period shift.

JANES: That is quite a large shift.

RENZINI: In fact, they say they cannot accept such a large change in the opacity.

SWEIGART: We have tried a number of numerical experiments to reproduce the observed period shift. We tried 1) including the effects of evolution away from the ZAHB, 2) changing the neutrino emission rate and conductive opacity during the red-giant phase and 3) enhancing the CNO abundance. None of these attempts were successful. The only experiment that worked was the one based on an increased metal opacity in the ZAHB models, as just mentioned by Alvio.

JANES: You left as a completely free parameter on your models, the total mass.

SWEIGART: What we are looking at is the period shift at a given effective temperature. This constraint fixes the mass for each composition, so the mass is not a free parameter.

JANES: What I am thinking of is the general morphology of the horizontal branch. Whether the total mass depends on the metallicity in some fashion that we don't know yet. It may depend on the amount of mass loss.

ADELMAN: Let me make some observational comments. Heasley and Wolff (1985 Astrophys. J. 292, 589.) found from λ He I 4026 you can get the helium abundance to 15%. I have a letter from Sidney which indicates that 10% accuracy is possible for λ He I 4471. So I don't see any real problems for determining helium abundances unless we have diffusion processes draining it away. The metallicity, I think, is a far worse problem. To determine it to a factor of two, is pretty difficult. We need very high signal/noise data. But even then you are going to have problems with gf values, damping constants and model atmospheres.

JANES: Diffusion does not make enough of a change over enough of a star to alter these parameters.

LECKRONE: But it would affect the numbers that you derive.

ADELMAN: That is what is going to make hot horizontal-branch star abundances difficult to interpret. You won't be able to determine their primordial helium abundances to the desired accuracv. You can measure the abundances with certainty for 10,000 K objects for which the metals are not affected by diffusion.

HOT POST-AGB STARS IN GLOBULAR CLUSTERS AND ELSEWHERE

Alvio Renzini

Department of Astronomy, Bologna

ABSTRACT. Planned space observatories (like UIT and ST) are likely to detect (and study) a significant number of hot Post-AGB stars which are members of galactic globular clusters, the galactic bulge and the bulge of M 31. Current Post-AGB evolutionary models predict that of the order of 100 such stars, brighter than 100 $L_{\odot}$, could be detected by surveying the 20 - 30 brightest galactic globular clusters. Some of the astrophysical motivations for undertaking a study of these stars are then briefly reviewed. These include: an accurate determination of the mass of currently forming white dwarfs in globular clusters, a detailed understanding of the Post-AGB stage of evolution in low mass stars and new insights concerning the Post-AGB contribution to the ultraviolet light of elliptical galaxies.

1. INTRODUCTION

Besides the blue horizontal-branch stars, globular clusters contain another kind of hot star; i.e. Post-AGB stars (AGB = Asymptotic Giant Branch). These objects are evolving towards their final WD stage, after having ejected most of their hydrogen-rich envelope (cf. Iben and Renzini 1983, for a recent review). There are several reasons why the study of globular cluster Post-AGB stars is particularly interesting. These include: (i) a determination of the final mass of currently dying globular cluster stars, (ii) an understanding of the final evolutionary stages of low mass stars, and (iii) an assessment of the contribution of Post-AGB stars to the ultraviolet output of old stellar populations.

2. SEARCHING FOR POST-AGB STARS IN GLOBULAR CLUSTERS

The bright portion of the Post-AGB phase of evolution has a rather short duration, and therefore Post-AGB stars are quite rare objects. When planning a search for such stars it is then most useful to have a way of guessing, *a priori*, how many of them can be found. This can be accomplished by using the concept of *specific evolutionary flux*, which is the number of stars entering (or leaving) any particular evolutionary phase, per year, and per unit luminosity ($L_{\odot}$) of the sampled stellar population (Renzini and Buzzoni 1985). In this way, the number of stars n_j in any given evolutionary stage is given by:

A. G. Davis Philip, (ed.)
Horizontal-Branch and UV-Bright Stars 19-29

$$n_j = B(t)\ L_T\ t_j, \qquad (1)$$

where B(t) is the specific flux, a function of the age, t, of the population, L_T is the luminosity of the sampled population, and t_j is the duration of the evolutionary stage under consideration. B(t) is displayed in Figure 1 as a function of age, and for three values of the slope of the initial mass function (IMF). Adopting t = 15 Gyrs for galactic globular clusters, Figure 1 gives B = 2 10^{-11} (stars per year per solar luminosity). Further, adopting L_T = 10^5 for the typical

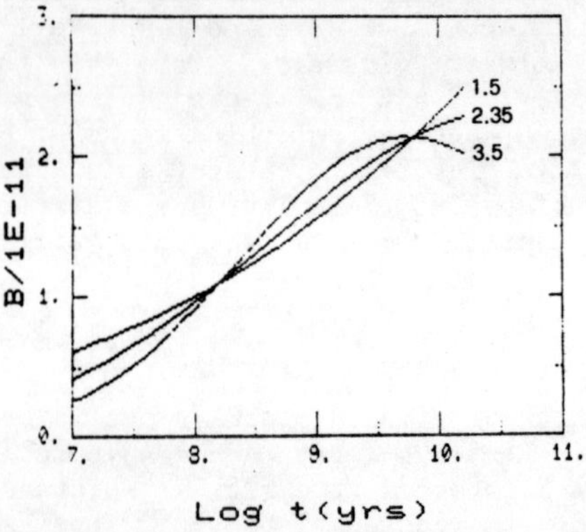

Fig. 1. The specific evolutionary flux B(t) as a function of age for three IMF slopes.

luminosity of a globular cluster, Eq. (1) then gives the expected number of Post-AGB stars, as a function of the time spent in this phase. Table I gives this number for a typical cluster, as well as for 100 such clusters, totalling 10^7 $L_\odot$. Therefore, in an assembly of clusters totalling 10^7 $L_\odot$ one might expect to find 100 Post-AGB stars which have left the AGB less than 5 10^5 years ago. This is a respectable number of objects, which should allow one to draw several meaningful conclusions.

TABLE I

Expected Number of Post-AGB Stars

T_{P-AGB}	$N_{P-AGB}10^5$	$N_{P-AGB}10^7$
10^4	0.02	2
10^5	0.2	20
5 10^5	1	100

The crucial question, however, concerns the luminosity of these stars. Figure 2 displays the Post-AGB evolutionary tracks constructed by Schönberner (1983), together with the temperatures and luminosities of known Post-AGB stars in globular clusters (from de Boer 1985). Several instructive considerations can be made on the basis of Figure 2. Namely: (i) the time required to fade down to a certain luminosity is an extremely sensitive function of the mass of Post-AGB stars, (ii) known Post-AGB stars are cooler than 35,000 K, while tracks extend to 100,000 K, and (iii) de Boer's data points indicate a typical mass around 0.55 $M_{\odot}$. This latter point is discussed in Section 3.

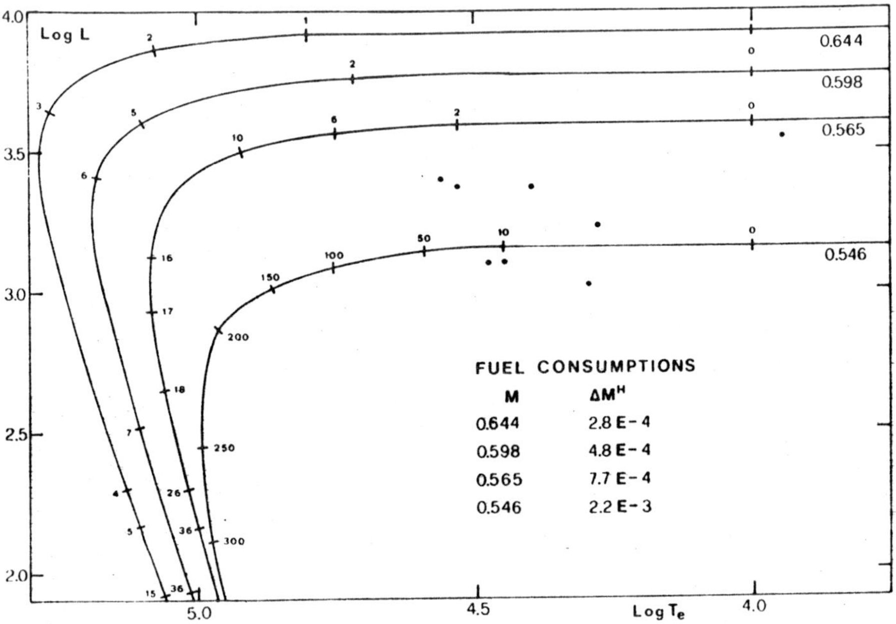

Fig. 2. The post-AGB tracks of Schönberner (1983), for the indicated post-AGB masses. Time marks (in 1000 years) give the time elapsed since models reach 10,000 K. Filled circles show the location of known post-AGB stars in galactic globular clusters (from de Boer 1985). In the insert, the corresponding hydrogen fuel consumptions in $M_{\odot}$ units.

Concerning point (i), the physical reasons for this behavior of the models are thoroughly discussed and explained elsewhere (Renzini 1983, Iben and Renzini 1983). Concerning (ii), one can note that all but one of the stars shown in Figure 2 were discovered through optical observations, which would have easily missed bolometrically bright hot stars, with very large bolometric corrections. Thus it is not surprising if known Post-AGB stars are found only at the cool end of the theoretically expected distribution. The exception is the UVB star in M 5, discovered by Bohlin et al. (1983) through UV imaging of the cluster with a prototype of the Ultraviolet Imaging Telescope (UIT, cf. Stecher et al. 1978). Not surprisingly this star is one of the hottest in de Boer's sample. The hottest star shown in Figure 2 is K 648 in M 15, which certainly would have been missed by optical surveys, were not for the fact that it is the central star of the homonymous planetary nebula.

If we now suppose that the eight stars in Figure 2 represent the complete sample of globular cluster Post-AGB stars that are cooler than 35,000 K, and that the total luminosity of the surveyed clusters is 10^7, then Eq. (1) allows us to estimate the duration of the portion of the Post-AGB phase which was actually sampled. Indeed, from $8 = 2 \times 10^{-11}\ 10^7\ t_{P\text{-}AGB}$, one gets $t_{P\text{-}AGB} = 40{,}000$ years, which is in rough agreement with the time one would derive from the 0.546 $M_\odot$ track in Figure 2. Moreover, this model fades to 100 $L_\odot$ in a 10 times longer time, and then one would expect to find 80 Post-AGB stars brighter than 100 $L_\odot$, in the same assembly of clusters. From Table I we also see that one would expect just 1 such star per cluster (on average), and then many clusters should be surveyed, in order to recover a statistically meaningful sample. One needs a wide-field instrument to carry out this type of survey, and the UIT is most appropriate. However, while UIT is the ideal instrument to pick these stars up, it can hardly give all the information which is required (effective temperature, bolometric luminosity, surface gravity and composition). The problem is that near-UV colors can only give a lower limit to the temperatures, for stars hotter than, say, 30,000 K. To sort out these quantities, a spectroscopic follow up is required, which could be accomplished with either HRS or FOS on board the Space Telescope.

3. DETERMINING THE MASS OF CURRENTLY FORMING WHITE DWARFS IN GLOBULAR CLUSTERS

Red Giant Branch (RGB) models indicate that the core helium flash takes place when the helium core has reached 0.50 $M_\odot$ (Sweigart and Gross 1978), and recent calculations show that this value is fairly insensitive to possible uncertainties in the input physics (Sweigart et al. 1985). Actually, a check of this theoretical prediction can be made by looking at the luminosity of the RGB tip in globular clusters, as this luminosity is fairly sensitive to the core mass at the helium flash. Frogel et al. (1983) did this comparison, finding fairly good agreement between predicted and observed RGB tip luminosities. A further check for the theoretically predicted He core mass comes from the observation of horizontal-branch (HB) stars, as the luminosity of HB models is sensitive to the He core mass (e.g. Sweigart and Gross 1976 and Sweigart 1985). It appears that current models do not predict the correct size for for the so-called Sandage period-shift effect, but nonetheless the differences are so subtle that it is fair to conclude that current theoretical estimates of the helium core mass cannot be wrong by more than a few 10^{-2}. We then have a doubly-checked lower limit for the mass of currently forming WD's in globular clusters, this limit being 0.50 $M_{\odot}$.

During the HB phase, the hydrogen-burning shell moves outwards in mass, thus increasing the core mass by an amount which is sensitive to both the stellar composition and mass (Sweigart and Gross 1976). This increase is typically of the order of 0.01 - 0.02 $M_{\odot}$. Following core helium exhaustion, stars enter the so-called Early-AGB phase, during which the hydrogen-burning shell is extinguished, while the helium-burning shell moves gradually outwards in mass. Correspondingly, the star climbs along the AGB, with a constant core mass, until the helium shell gets very close in mass to the hydrogen shell. At this point the hydrogen shell is reignited and shortly afterwards the star experiences its first helium shell flash, thus entering the Thermally Pulsing AGB phase (Iben and Renzini 1983). This happens when the star has a surface luminosity slightly in excess of 10^{3} $L_{\odot}$, i.e. very close to that of the RGB tip. Subsequently, the (average) stellar luminosity increases in pace with the increase in core mass, following the luminosity-core mass relation first discovered by Paczynski (1970). Therefore, a determination of the luminosity of the AGB termination provides an estimate for the final value of the core mass, and then for the mass of currently forming WD's, M_{WD}.

Apart from the most metal-rich clusters, in globular clusters the AGB never extends beyond the RGB tip, and then from the luminosity-core mass relation one gets an upper limit for the final core mass, and Fusi Pecci and Renzini (1979) estimate M_{WD} < 0.54 $M_{\odot}$. In conclusion, when putting together the lower and upper limits, one should have 0.50 < $M_{\odot}$ < 0.54. As pointed out by Fusi Pecci and Renzini, such a fairly precise estimate of M_{WD} may be of great value for a better determination of the distance to globular clusters, and then of their age (see also Iben and Renzini 1984).

For these reasons any additional check for the predicted value of M_{WD} is particularly important, and indeed Post-AGB stars may be very useful in this respect. Following envelope ejection, stars evolve at almost constant luminosity towards higher and higher temperatures, as shown in Figure 2. While on the horizontal portion of the Post-AGB track, the stellar luminosity is linked to mass through the same core-mass luminosity relation holding for AGB stars. The only difference is that the envelope mass is now very small (< few 10^{-3} $M_{\odot}$), and then the core mass is practically equal to the total stellar mass. Therefore, mass can be obtained from the observed luminosity, for stars lying on the horizontal portion of their Post-AGB track, and indeed from Figure 2 one can estimate an average mass of around 0.54 - 0.55 $M_{\odot}$. Certainly, an even better determination of M_{WD} will be possible, once many more Post-AGB stars could be added to the diagram. It is worth emphasizing here the remarkable accuracy of this weighting method, as the Post-AGB luminosity is so sensitive to mass. In principle, an accuracy of ± 0.01 $M_\odot$ is possible, certainly superior to any other weighting method, such as, for instance, through spectroscopically estimated surface gravities.

Finally, one may point out that the brightest (and coolest) star in Figure 2 belongs to 47 Tuc, whose AGB stars do actually extend to a higher luminosity compared to other, more metal-poor clusters (Frogel 1983). So AGB and Post-AGB stars consistently give a hint that M_{WD} may slightly increase with increasing metallicity. Again, this possibility could be checked in greater detail, once other Post-AGB stars are detected and studied.

4. THE POST-AGB EVOLUTION OF LOW MASS STARS

In this section I will briefly mention new insights we may gather on the Post-AGB evolution of globular cluster stars, once some 100 such stars could be identified, spectroscopically studied, and then placed in the HR diagram. These new insights may include: (i) checking the time spent in different portions of the Post-AGB track, (ii) assessing whether mass loss has any influence on the rate of evolution, (iii) establishing which fraction of Post-AGB stars experiences a final helium shell flash and develops a hydrogen-free surface, and (iv) comparing surface abundances to theoretical predictions. Point (i) does not require much further illustration; simply the number distribution of stars along the Post-AGB sequences ought to be compared with the corresponding evolutionary lifetimes. This may actually lead to an even better determination of M_{WD}, were it not for possible complications brought about by point (ii). Indeed, during the horizontal portion of the Post-AGB track, the rate of evolution is controlled by the rate of decrease in the mass of the residual, tiny hydrogen-rich envelope. If Post-AGB stars are losing mass at a rate comparable to the rate at which hydrogen is being burned, then the evolution can be accelerated significantly.

For instance, in the case of the 0.546 $M_\odot$ model, hydrogen burning reduces the envelope mass at a rate of 2.2×10^{-8} $M_\odot$/yr, and correspondingly mass loss rates of this order will greatly accelerate the rate of evolution. Current estimates of the mass loss rate from the central stars of planetary nebulae range from 10^{-10} to 10^{-7} $M_\odot$/yr (Cerruti-Sola and Perinotto 1985), and are individually uncertain by perhaps an order of magnitude, or more. Adams et al. (1984) estimate 2×10^{-10} $M_\odot$/yr for the central star of K 648 in M 15, some two orders of magnitude lower than the burning rate, but all these estimates are too crude, anyway, for assessing whether mass loss affects the rate of evolution. Conversely, the number distribution of stars along the Post-AGB sequence can give more cogent information about the role of mass loss.

Concerning point (iii), theory would predict that perhaps 20% of Post-AGB stars may experience a final helium shell flash, and it has been argued that under such circumstances the star is likely to expose intershell material, thus turning into a hydrogen deficient star, appearing first as a Wolf-Rayet type planetary nebula nucleus, and later evolving to a non-DA white dwarf (Renzini 1979, 1982, 1983, Iben et al. 1983, Iben and Renzini 1983 and Iben 1984). The UV spectroscopic follow up (with ST) of Post-AGB stars discovered with the UIT surveys should then allow one to check in great detail this theoretical prediction, therefore assessing which fraction of Post-AGB stars actually do experience a final helium shell flash. Moreover, this fraction could be compared meaningfully to the non-DA/DA ratio of white dwarfs in globular clusters that ST observations are likely to determine. Moreover, stars in globular clusters depart from the AGB at luminosities which are very close to that at which they are expected to experience their first helium shell flash, and then it may well be that such a flash takes place while some stars have already departed from the AGB, and have reached high surface temperatures. In such a case this would be the only He shell flash actually experienced, an event which has never been numerically explored, and which may hold in reserve some surprise.

Finally, concerning point (iv), IUE observations of the few known halo planetary nebulae indicate large carbon overabundances (Torres-Piembert 1984, Adams et al. 1984), which are not predicted by current theoretical models. Elemental analysis of Post-AGB stars in clusters would shed additional light on what is actually going on during the latest evolutionary stages of low mass stars.

In concluding this section, it is worth mentioning that any trend in Post-AGB characteristics with metal abundance could be investigated fruitfully, if the number of stars studied were to be large enough (say, 100 stars or so).

5. THE POST-AGB CONTRIBUTIONS TO THE UV LIGHT OF OLD STELLAR POPULATIONS

The spectra of elliptical galaxies show an upturn shortwards of 2000 Å, indicating the presence of very hot stars (cf. for instance, Faber 1983 and references therein). The nature of these hot stars has not been established yet, and the various possibilities still include: young massive stars, hot horizontal-branch stars, Post-AGB stars and binaries (cf. Renzini and Buzzoni 1985, for an extensive discussion). I will not repeat this discussion here but just mention how the detection of Post-AGB stars in clusters may help in a better understanding of the origin of UV light in elliptical galaxies.

From a theoretical point of view, the overall UV luminosity, L_{UV}, provided by Post-AGB stars in a given stellar population is proportional to the amount of fuel which is burned during this phase. In turn, this fuel consumption is extremely sensitive to the actual mass of Post-AGB stars, as one can appreciate from the table inserted in Figure 2. Therefore, the estimated fractional contribution of Post-AGB stars (L_{UV}/L_T) is extremely sensitive to the adopted mass of Post-AGB stars. Here L_T, as in Eq. (1), is the total bolometric luminosity of the stellar population. The actual detection of a complete sample of hot Post-AGB stars in galactic globular clusters will allow us to estimate empirically their total L_{UV} (= Σ L_i, where the summation is extended to all detected Post-AGB stars), and then to evaluate the L_{UV}/L_T ratio in the system of galactic globular clusters. Here L_T is the total bolometric luminosity of all surveyed clusters, including those in which Post-AGB stars are absent. This global approach is required to cope with small number statistics, which will completely dominate the L_{UV}/L_T ratio for individual clusters. Provided a sufficient number of clusters will be surveyed, it might be possible to bin clusters in a few chemically homogeneous groups, therefore testing whether there is any trend of L_{UV}/L_T with metallicity.

This latter point introduces the important question of the L_{UV}/L_T ratio in the old, high metallicity populations. Indeed, the bulk of stars in elliptical galaxies is probably metal rich, and it may be unsafe to extrapolate, by perhaps over one dex in [Fe/H], any trend in L_{UV}/L_T vs [Fe/H] which might have been established for galactic globular clusters. The observation of individual Post-AGB stars in an old, metal-rich environment is therefore crucially important. UIT imaging of the galactic bulge through the Baade window may help considerably, although the interpretation of the results will be complicated by reddening effects. FOC observations in the bulge of M 31 may prove more fruitful in this context.

6. CONCLUSIONS

In this paper I just aimed at providing some flavor for the kind of things that will soon become feasible for Post-AGB stars. This listing has certainly been incomplete, and superficial, and I am aware that the UIT team has already planned most of the observations which are discussed here. I hope however that this cursory review may have given some useful indications, and I conclude by emphasizing that most of our theoretical understanding of the Post-AGB stage of evolution still relies on a very limited number of theoretical evolutionary sequences. This stage is actually so sensitive to mass and other fine details, and exhibits such a wonderful variety of behavior, that an extensive exploration of the parameter space would be most urgent and productive.

ACKNOWLEDGEMENTS

I would like to thank R. Bohlin, R. O'Connell and T. Stecher for very useful conversations. I am also grateful to the Space Telescope Science Institute for its hospitality during the period in which this paper has been written and set up. Finally, I thank Ms. D. Schlogel for her patient and competent editorial cooperation.

REFERENCES

Adams, S., Seaton, M. J., Howarth, I. D., Auriere, M, and Selby, M. J. 1984 Monthly Notices Roy. Astron. Soc. 207, 471.
Bohlin, R. C., Cornett, R. H., Hill, J. K., Smith, A. M., Stecher, T. P. and Sweigart, A. V. 1983 Astrophys. J. Lett. 267, L89.
Cerruti-Sola, M. and Perinotto, M. 1985 Astrophys. J. 291, 237.
de Boer, K. S. 1985 Astron. Astrophys. 142, 321.
Faber, S. M. 1983 Highlights of Astronomy 6, 165.
Frogel, J. A. 1983 Astrophys. J. 272, 167.
Frogel, J. A., Cohen, J. C. and Persson, S. E. 1983 Astrophys. J. 275, 773.
Fusi Pecci, F. and Renzini, A. 1979 Astronomical Uses of the Space Telescope, F. Macchetto, F. Pacini and M. Tarenghi, eds., (ESO) p. 181.
Iben, I. Jr. 1984 Astrophys. J. 277, 333.
Iben, I. Jr., Kaler, J. B., Truran, J. W. and Renzini, A. 1983 Astrophys. J. 264, 605.
Iben, I. Jr. and Renzini, A. 1983 Ann. Rev. Astron. Astrophys. 21, 271.
Iben, I. Jr. and Renzini, A. 1984 Physics Reports 105, 329.
Paczynski, B. 1970 Acta Astronomica 20, 47.
Renzini, A. 1979 in Stars and Star Systems, B. E. Westerlund, ed., Reidel, Dordrecht, p. 155.
Renzini, A. 1982 in Wolf-Rayet Stars, C. de Loore and A. J. Willis, eds., Reidel, Dordrecht, p. 413.
Renzini, A. 1983 in Planetary Nebulae, D. R. Flowers, ed., Reidel, Dordrecht, p. 267.
Renzini, A. 1985 in The Spectral Evolution of Galaxies, C. Chiosi and A. Renzini, eds., Reidel, Dordrecht, in press.
Schonberner, D. 1983 Astrophys. J. 272, 708.
Stecher, T. P. 1978 NASA UIT-680-78-01.
Sweigart, A. V. 1985 in Horizontal-Branch and UV-Bright Stars, A. G. D. Philip, ed., L. Davis Press, Schenectady, p. 3.
Sweigart, A. V. and Gross, P. C. 1978 Astrophys J. Suppl. 36, 405.
Sweigart, A. V., Renzini, A. and Tornambe, A. 1985 In preparation.
Torres-Piembert, S. 1984 in Stellar Nucleosynthesis, C. Chiosi and A. Renzini, eds., Reidel, Dordrecht, p. 3.

DISCUSSION

JANES: For these calculations are metallicities an important parameter?

RENZINI: It has not really been explored.

PHILIP: With the UIT satellite that you mentioned, what is the size of the field and how much time would you need per exposure?

RENZINI: I think that they would be quite short. They are extremely bright; they are among the brightest stars in the cluster, of the order of 1000 solar luminosities.

PHILIP: Is the field large enough to do one globular cluster in one exposure?

RENZINI: I think that you get the whole cluster at once.

JANES: I think that it is a 40 minute of arc field.

RENZINI: Presumably the exposures would be short, a few minutes. Then you would have a natural followup, to do spectroscopy on the stars that are found.

JANES: Is it that clear that the present observations of clusters would not have picked up such stars?

RENZINI: At 60,000 to 100,000 degrees I think that the bolometric correction is so large that they would have been missed. Certainly, for those that are in the central part of the cluster there is a crowding problem. But making a picture in the UV will help pick up these stars. You will see only the blue horizontal-branch stars and the uv-bright stars.

JANES: With the deep color-magnitude surveys in the outer parts of clusters the problem is even worse.

ADELMAN: You can always take short exposures to locate the bright UV stars in the center of the cluster. One mission will take about 200 good exposures.

RENZINI: To do a complete survey of globular clusters will take about 200 exposures.

LECKRONE: We said last time that these stars should be very highly mixed. So if you could measure compositions spectroscopically -

SWEIGART: Since they have recently been over the AGB they would have had deep convective envelopes and thus would be thoroughly mixed. The important question is how quickly does diffusion work.

ADELMAN: Since the known stars have lifetimes of less than 10^4 years in their current stage of evolution diffusion is not expected to be important. The hotter stars in this evolutionary sequence are the ones for which you would have to worry about such effects. Once you get hotter than 30,000 K, I really don't want to worry about what the He I lines are like anyhow.

LECKRONE: That's the problem, as you get to the higher temperatures the spectroscopic diagnostics get more complicated.

RENZINI: There are many other interesting aspects connected to composition. For instance, the planetary nebula in M 15, which was one of those points, is extremely rich in carbon which suggests a dredge-up of carbon. Once again, the models do not predict a third dredge-up for such small mass stars. One may also predict that 10 - 20 % of these stars to be in a phase following the final helium shell flash, while the star retraces essentially the same evolutionary line and loses the hydrogen rich envelope, and should show up as a Wolf-Rayet star of low mass, like some nuclei of planetary nebulae. So it will be interesting to see, out of this complete sample of post AGB stars, what is the fraction of Wolf Rayet against normal stars. Once you have the sample you can do a lot of things with it.

RED GIANTS AND THE HORIZONTAL BRANCH

K. Janes

Boston University

ABSTRACT. The various processes which occur as a star evolves from the main sequence along the giant branch are directly relevant to an understanding of the character of the horizontal branch. The initial composition of stars in a globular cluster is most easily found from observing the giants, while the extent to which nuclear-processed material mixes to the surface and the amount of mass loss; all are red giant processes which are directly relevant to an understanding of the horizontal branch. On the other hand, some of the outstanding questions relating to the giant branch impact directly on our understanding of the evolution of globular clusters, and cannot be understood without an examination of the properties of horizontal-branch stars. The HB morphology, abundances of He, C, N and O and the pulsational masses of HB stars; all can help to untangle the course of evolution along the giant branch. Our present state of understanding of the globular cluster giant branch is reviewed.

1. INTRODUCTION

Why should a review of the globular cluster giant branch be relevant to a meeting on hot horizontal-branch stars? The answer to this question comes from stellar evolution considerations: horizontal-branch stars have, of course, evolved directly from the giant branch, and they in turn evolve back into red giants on the asymptotic giant branch. At the very end of their lives they become hot again briefly as they evolve into "UV-bright" stars on their way to becoming white dwarfs. The general picture is well-summarized in Figure 1, adapted from a diagram of Iben (1981) which shows not only the evolutionary track of a low mass star, but also the locations of several critical events, such as the first dredge-up phase.

The detailed evolutionary track of a star depends on its initial composition, but in fact, most of what is known about globular cluster compositions has been derived from observations of red giant stars. Unfortunately, the globular cluster abundance scale is still a matter of some controversy, and there is conflicting evidence as to whether observed abundance variations within a given cluster are due to effects

A. G. Davis Philip, (ed.)
Horizontal-Branch and UV-Bright Stars 31-39

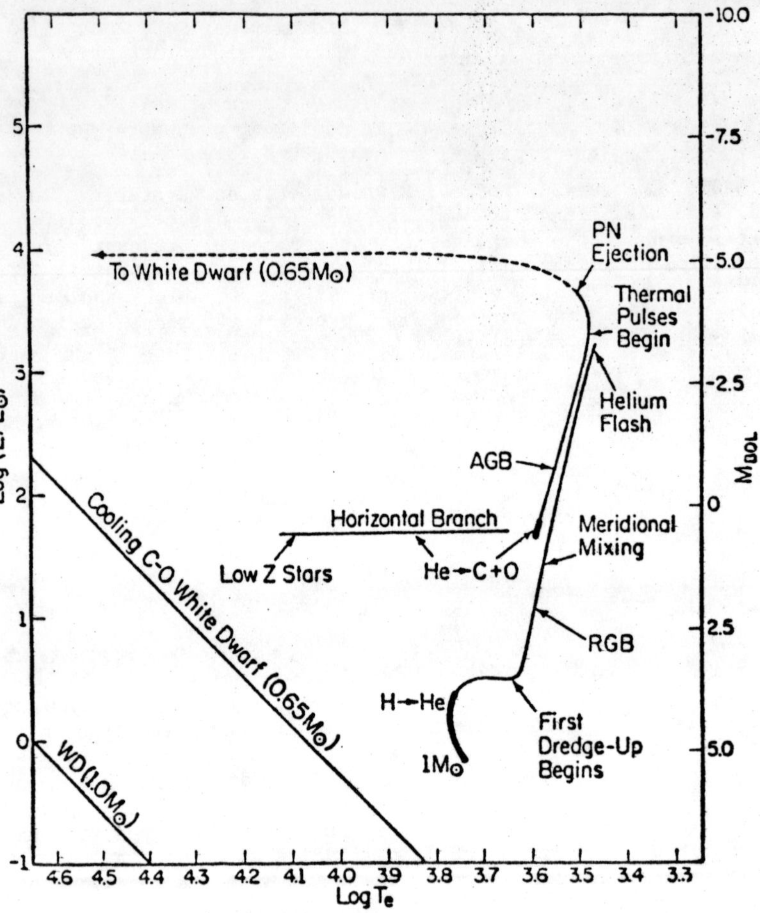

Fig. 1. Schematic evolutionary track of a one solar mass star, from the main sequence to the white dwarf stage. Adapted from a diagram of Iben (1981).

as individual stars evolve or whether they are the result of primordial variations (see reviews by Kraft 1979 and Freeman and Norris 1981). Next, there are a lot of direct and indirect signs that mixing in red giant interiors brings nuclear-processed material to the surface, most likely either at the first dredge-up phase or as a result of meridional mixing (Iben 1981, see Figure 1). In addition, somewhere between the main sequence and the horizontal branch, stars lose about a quarter of a solar mass, either catastrophically (possibly at the time of the helium flash) or more likely gradually in the form of a stellar wind (Reimers 1981). Finally, we have very little direct knowledge of the abundances of the important elements He, C, N and O. In this case, studies of the horizontal branch will contribute to understanding the giant branch, and to cluster evolution in general. The purpose of this review is to recapitulate some of the important aspects of stellar evolution and the properties of globular cluster red giants as they relate to the problems of the hot stars in the globular clusters.

2. THE GLOBULAR CLUSTER ABUNDANCE SCALE

Most of what we know about the compositions of globular clusters comes from studies of the giants. Because all globular cluster stars are faint, observations at high dispersion are difficult and most measurements of their compositions are made by calibrating low resolution spectra and photometry using existing high dispersion spectra. Unfortunately, there has been a surprising lack of agreement as to the scale of globular cluster abundances based on such calibrations. The principal reason for this difficulty is that until very recently, the generally accepted "old" scale of globular clusters abundances was based primarily on high dispersion spectra of just two stars, one in M 92 (Helfer et al. 1959) and one in M 13 (Pagel 1966). Although Butler (1975) obtained an independent measure, by using the delta-s method of low-resolution spectra to measure RR Lyrae stars in 12 clusters, his work was calibrated with curve-of-growth analyses of field RR Lyraes. For the most part, the two decade gap between the compositions of these two stars and normal Population I stars has been bridged by interpolation, using various types of photometry and low-resolution spectroscopy. The uncertainties are most severe among the more metal-rich globular clusters because they are in the middle of the large gap in composition; on this old scale even the "metal-rich" clusters, M 71 and 47 Tuc have [Fe/H] = -0.35 (Harris and Racine 1970) making them more metal poor than the great majority of Pop I stars.

The advent of echelle spectrographs on 4-meter-class telescopes in the past ten years has permitted a dramatic change in this situation. Pilachowski (1984, and other references therein), Cohen (1979, 1981) and Peterson (1980) have gathered high-dispersion observations of stars in 23 clusters. The "new" scale resulting from their work is significantly different from the old one at the metal-rich end where there were no direct measures before; for example Pilachowski finds [Fe/H] = -1.2 for the metal-rich clusters 47 Tuc and M 71, a dramatic change. On the other hand, the most metal-poor clusters have not changed significantly from the older scale.

Because of the much larger data base, the new scale would appear to be on a significantly better footing than the old; however there are some disquieting consequences. Primarily, there is no longer a simple correlation between [Fe/H] and the various indices of photometry and low-resolution spectroscopy or with the color-magnitude diagram morphology (see Zinn and West 1984), and in fact, the correlation even disappears at the metal-rich end. While this does not, by itself, prove that there is something wrong with the new scale, it is a cause for some suspicion and would require some sort of explanation as to why a diverse set of photometric indices is uniformly discordant with the spectroscopy. Furthermore, Cohen (1981) included the old open cluster, M 67, in her program and found [Fe/H] = -0.3, whereas most other observers find that M 67 is about [Fe/H] = 0.0 (see Janes 1985, for a review of the properties of M 67).

This change in globular cluster abundances, if verified, would require radical alterations in our view of the history and nature of the globular clusters. Consequently, the new results have generated a great deal of interest, tempered with some skepticism, and several re-examinations of the new data have been made.

Pilachowski (1984, see also Pilachowski et al. 1983) reanalysed all existing high-dispersion work using a uniform technique and compared her analysis with several photometric indices. She concluded that while the high-dispersion data consistently gives low values for the M 71 and 47 Tuc metallicities, other methods just as consistently give higher values. Furthermore, the question of the metal-rich end of the scale still rests almost entirely on those two clusters. She therefore adopts [Fe/H] = -0.8 as a compromise value, and develops a cluster metallicity scale on that basis.

In another effort to resolve this difficulty, Cohen (1983) observed several more metal-rich clusters at moderate resolution and re-analysed 47 Tuc with new, high signal-to-noise spectra. This gave her a new scale of metal-rich clusters relative to 47 Tuc and for the latter cluster a value of [Fe/H] = -0.7. Frogel et al. (1983) used these data with older spectroscopic values for metal-poor clusters to set up a list of 16 globular clusters to use for metallicity calibrations.

About the same time, Bell and Gustafsson (1983) did a spectrum synthesis of low-resolution spectra and concluded that the higher abundances are necessary. The Bell-Gustaffson revision of the new scale is basically in agreement with Cohen's; on their scale, the 47 Tuc/M 71 metallicity is of the order of [Fe/H] = -0.7 or slightly less.

Zinn (1980a, 1980b, 1985), Zinn and West (1984), in a series of four papers produced a comprehensive set of observations and did a painstaking analysis of the entire question of globular cluster metallicities. Using an integrated light measure of globular clusters based on the line blanketing between 3820 - 4010 A, he formed an index, Q39, which he used to determine the metallicities of over 100 clusters. When he compared the Q39 index to a number of other metallicity indices based on low-resolution spectroscopy, photometry and the morphology of the color-magnitude diagram he found in all cases an excellent correlation. Whatever it is that these indices are all measuring, it is the same thing in each case. Because all of the indices are so well correlated with one another, he was able to combine them into a single metallicity estimate, on a self-consistent system, for 121 clusters. For many of the clusters, with more than one type of measure, the error, within the system, is very small (frequently less than 0.1 in [Fe/H]). But when he tried to relate Q39 to the "new" system of abundances, he found not only a non-linear relation, but at the metal-rich end, even a lack of any real correlation at all. The revised Cohen scale, however, can be related directly to Q39 (and hence to other indices as well). Zinn's recent work is transformed into this system.

Using the revised new scale (as formalized by Frogel et al. 1983), a monotonic, nearly linear relation between the spectroscopic abundances and low-resolution composition indices is restored. Thus Zinn's (1985) table of metallicity data for globular clusters is of unprecedented homogeneity and precision; although further surprises are always possible, the weight of the evidence is that the revised new scale and Zinn's catalog are unlikely to be more in error than perhaps 0.2 in [Fe/H].

3 MIXING AND MASS LOSS ALONG THE GIANT BRANCH

The evidence that stars lose mass and mix their interior material to the surface is pervasive, but indirect and difficult to quantify. Among Population I stars, the first "dredge-up" phase is fairly well understood, both theoretically and observationally (Lambert 1981). This phase, occurring as a star begins its first ascent of the giant branch (see Figure 1) results when the convective core descends far enough into the interior of the star to make contact with material that has experienced partial CN-processing in the main sequence phase. As a result of this episode, we expect to see a decrease in the $^{12}C/^{13}C$ ratio from about the solar value of 89, a depletion of the fragile nuclei, Li, Be and B, a decrease in ^{12}C, an increase in ^{14}N, but no change in ^{16}O. All of these predicted changes are observed in about the right amounts in nearby Pop I giants, which can be observed easily at high dispersion (see the review by Lambert 1981).

In the Asymptotic Giant Branch (AGB) phase, there may be a second dredge-up phase, with a significant increase in He, and a further decrease in ^{12}C; but this is probably active only in stars between about 3 - 8 solar masses. A third dredge-up may result from H and He double-shell instabilities, producing carbon stars, as well as s-process enhancements (Iben 1984).

Presumably some of the same processes occur among Pop II stars as well, but the observations are much less complete, and the picture is complicated by questions concerning abundance variations among stars within a single cluster. As described by Kraft (1979) and Freeman and Norris (1981), the abundance variations may be primordial, or the variations may occur as a star evolves. In either case it will be necessary to study the mixing question in more detail; given the known range of compositions within a cluster, it is not enough to say that whatever happens to one star at a given phase of evolution will happen to all of them. The second dredge-up phase occurs only in relatively massive stars, so it is not seen in globular clusters. The third dredge-up also depends on mass as well as composition, and perhaps could occur in stars as low in mass as the globular cluster stars. There may be observed s-process enhancements, which could be indicative of the third dredge-up, but carbon stars are not seen in globular clusters. It is probable that Population II stars are too low in mass for the third dredge-up to be operative (Iben 1984).

The possibility also exists that meridional mixing may alter the surface compositions of globular cluster red giants. As reviewed by Iben (1984), there is evidence for decreasing carbon abundance in M 92 stars with increasing luminosity, continuing to luminosities above which the first dredge-up would have been expected to have been operative. A possible explanation for this phenomenon is that in the densely packed star cluster, encounters among stars might have induced temporary spin-ups of some stars, resulting in meridional mixing.

All stars lose mass; in the Sun, the amount of mass loss through the solar wind is less than 10^{-10} solar masses per year, but the rate increases with decreasing gravity and temperature in a more-or-less continuous fashion. Reimers (1981) has surveyed the observational evidence for mass loss as detected in circumstellar (CS) lines. He finds that all M giants and supergiants show optical CS lines, some G and K supergiants do and a few K giants show them, but stars that are hotter or less luminous than these do not. These CS lines show a characteristic P Cygni profile, with velocities much less than, but correlated with, the escape velocity. In the ultraviolet, there is a rather sharp dividing line between stars with transition layer lines (i.e., those of N V, V IV, He II, etc.) and those with low-temperature chromospheric lines, such as those of OI, C I or Si II. The former are indicative of hot, stationary coronae, whereas the latter have cool, high-velocity (about 100 km/sec) winds.

The transition between the hot coronae and cool winds occurs at roughly the same types of stars as the transition between stars without and with CS lines. The implication is that the amount of mass loss is relatively small as a star evolves upward along the giant branch, until it reaches the boundary, where there is a fairly sudden onset of relatively rapid mass loss. It is uncertain whether Population II stars reach sufficient luminosities for the rapid mass loss to occur. Cohen (1976) has observed H-Alpha emission in globular cluster red giants, but

Reimers (1981) argues that these particular lines are not indicative of significant mass loss. Mass loss does occur among Population II stars, because the horizontal-branch stars are less massive than their main sequence progenitors (by about 0.25 solar masses), but the process itself has not definitively been detected.

4. THE MORPHOLOGY OF THE COLOR-MAGNITUDE DIAGRAM

The most direct link between stellar evolution models and observed globular clusters comes from a comparison of the observed and theoretical color-magnitude diagrams. Among the many points that could be mentioned, perhaps the most relevant to the present discussion is the so-called "second parameter effect". Among clusters presumed to be similar in composition (as determined from observations of the red giants!), some have horizontal-branch stars with mostly blue stars and some with only red horizontal-branch stars. Among clusters in the outer halo, at a given metallicity, there is a wide range of horizontal-branch types, but in the inner halo, all clusters of a given metallicity resemble each other. The basic effect can be seen in Figure 9 of Demarque (1980). In the inner halo, a metal-poor cluster will have a blue horizontal branch, whereas a metal-rich one will have a red one. Only in a relatively small transition region at intermediate metallicity will the horizontal branch contain both red and blue stars, but that just happens to occur at a common value of metallicity. In the outer halo, many rather metal-poor clusters have red horizontal branches.

This phenomenon has been known for some time and has generated a great deal of speculation as to its explanation. (See Kraft 1979 and Freeman and Norris 1981 for example.) At the transition metallicity between a red and a blue horizontal branch, the detailed morphology is extremely sensitive to several possible parameters, such as age, rotation, He, C, N and O abundances. Consequently, in the outer halo, at least one of these parameters must vary independently of metallicity, but in the inner halo, no such independence is detected. Although strong opinions have been expressed, at the moment, any one of these parameters is a viable candidate for the second parameter.

5. SUMMARY

We find, therefore, that our picture of the character of horizontal-branch stars is very much colored by what we know about red giants, and vice versa. The most critical problem is the globular cluster metallicity scale, which so far has been determined almost entirely from observing red giants. Until we can be sure that the new metallicity scale, as given by Frogel et al. (1983) is reliable, then almost anything one attempts to say about globular clusters will be only tentative. Furthermore, the star-to-star variations in composition, which are seen among the red giants, also must be present on the horizontal branch, affecting the morphology of the horizontal branch in an unknown fashion. On the other hand, to find the abundances of He, C, N and O, which affects the course of stellar evolution, observations of horizontal-branch stars are required. Finally, in order to learn about the processes of mixing and mass loss, a combined study of both red giants and horizontal-branch stars will be needed. The amount of mass loss of course directly affects the character of the horizontal branch.

REFERENCES

Bell, R. A. and Gustaffson, B. 1983 Monthly Notices Roy. Astron. Soc. 204, 249.

Butler, D. 1975 Astrophys. J. 200, 68.

Cohen, J. G. 1976 Astrophys. J. 203, 210.

Cohen, J. G. 1979 Astrophys. J. 231, 751.

Cohen, J. G. 1981 Astrophys. J. 247, 869.

Cohen, J. G. 1983 Astrophys. J. 270, 654.

Demarque, P. 1980 in IAU Symposium No. 85, Star Clusters, J. E. Hesser, ed., Reidel, Dordrecht, p. 281.

Freeman, K. C. and Norris, J. 1981 Ann. Rev. Astron. Astrophys. 19, 319.

Frogel, J. A., Cohen, J. G. and Perrson, S. E. 1983 Astrophys. J. 275, 773.

Harris, W. E. and Racine, R. 1979 Ann. Rev. Astron. Astrophys. 17, 241.

Helfer, J. L., Wallerstein, G. and Greenstein, J. L. 1959 Astrophys. J. 129, 700.

Iben, I. 1981 in Physical Processes in Red Giants, I. Iben and A. Renzini, eds., Reidel, Dordrecht, p. 3.

Iben, I. 1984 in IAU Symposium No. 105, Observational Tests of the Stellar Evolution Theory, A. Maeder and A. Renzini, eds., Reidel, Dordrecht, p. 3.

Janes, K. A. 1985 in IAU Symposium No. 111, Calibration of Fundamental Stellar Quantities, D. S. Hayes, L. E. Pasinetti and A. G. D. Philip, eds., Reidel, Dordrecht, p. 361.

Kraft, R. P. 1979 Ann. Rev. Astron. Astrophys. 17, 309.

Lambert, D. L. 1981 in Physical Processes in Red Giants, I. Iben and A. Renzini, eds., Reidel, Dordrecht, p. 115.

Pagel, B. E. J. 1966 in Colloquium on Late Type Stars, M. Hack, ed., Observatorio Astronomico, Trieste, p. 133.

Peterson, R. 1980 Astrophys. J. Lett. 237, L87.

Pilachowski, C. A. 1984 Astrophys. J. 281, 614.

Pilachowski, C. A., Sneden, C. and Wallerstein, G. 1983 Astrophys. J. Suppl. 52, 241.

Reimers, D. 1981 in Physical Processes in Red Giants, I. Iben and A. Renzini, eds., Reidel, Dordrecht, p. 269.

Zinn, R. 1980a Astrophys. J. Suppl. 42, 19.

Zinn, R. 1980b Astrophys. J. 241, 602.

Zinn, R. 1985 Astrophys. J. in press.

Zinn, R. and West, M. J. 1984 Astrophys. J. Suppl. 55, 45.

DISCUSSION

ADELMAN: C. W. Clark (1984 Astrophys. J. 285, 322.) calculates the isotopic shifts for some lines of light elements. You get a little rotation and the effects become non-observable.

THE IDENTIFICATION AND STUDY OF FIELD HORIZONTAL-BRANCH A-TYPE STARS

A. G. Davis Philip*

Van Vleck Observatory and Union College

ABSTRACT. By means of spectral plates obtained with Schmidt telescopes and follow-up photoelectric observations in the four-color system of all the A-type stars marked on the plates, a number of field horizontal-branch (FHB) A-type stars have been identified in areas at high galactic latitude. Data concerning other FHB stars have been taken from the literature. Various types of additional observations point out the similarity of the FHB to blue horizontal-branch (BHB) stars in globular clusters. This paper describes methods used to identify HB stars in the field and contains a summary of information known concerning FHB stars.

1. INTRODUCTION

In 1978 a summary article was written concerning "Horizontal-Branch Stars in the Log T_{eff}, Log g Diagram" (Philip 1978) in which references to four-color photometric measures of field horizontal-branch (FHB) and blue horizontal-branch (BHB) stars in globular clusters were listed. The four-color indices contained in these papers were plotted in various photometric diagrams. The HB stars being discussed are of the type that fall between (B-V) = 0.00 and 0.30 at V = 15.0, as shown in Fig. 1, a CM diagram of M 5 (a composite diagram from Philip, Cullen and White (1976) in which photometry from Arp (1962), Simoda and Tanikawa (1970) and Eggen (1972) were combined into one diagram). In the years since 1978, additional measures have been made in the four-color system of BHB and FHB stars and the precision of the calculated indices has increased due to the increased number of observations per star. Other observations have been made which tend to show that the field HB stars are very much like the BHB stars in globular clusters. The purpose of this paper is to summarize the old and new data, to indicate the methods by which FHB stars can be identified and to indicate what has been learned about FHB stars in the last decade.

---

* Visiting Astronomer Kitt Peak National Observatory and Cerro Tololo Inter-American Observatory, Mt. Palomar Observatory and Steward Observatory.

A. G. Davis Philip, (ed)
Horizontal-Branch and UV-Bright Stars 41-56

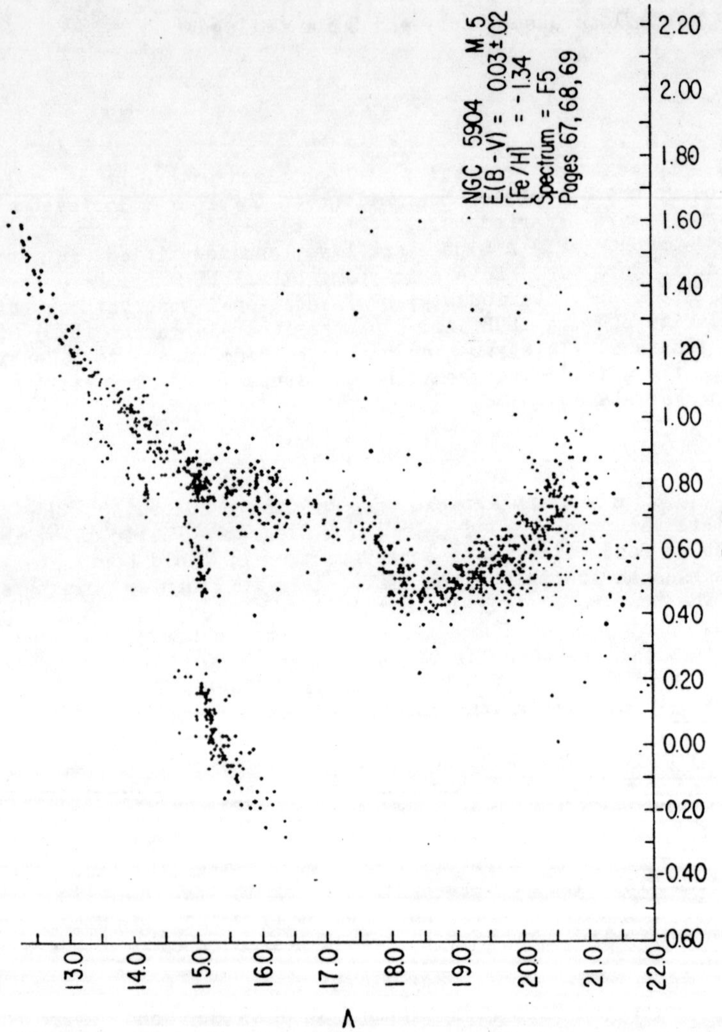

Fig. 1. Composite CM Diagram for NGC 5904 = M 5.

A few FHB stars were definitively identified as Population II objects by Oke, Greenstein and Gunn (1966) from photoelectric spectrum scans which were matched to stellar atmospheric models. The models that matched best were those that had the temperatures and surface gravities typical of horizontal-branch stars. In the 1940's and 50's surveys of high galactic latitude regions (such as that of Humason and Zwicky (1947), those made at Tonantzintla with the Schmidt telescope, and that of Feige (1958)) had turned up great numbers of faint blue stars and further photoelectric work identified various potentially interesting objects. Three of the stars in the Oke, Greenstein and Gunn list (HD 86986, 109995 and 161817) also appeared in a list of high velocity stars compiled by Roman (1955) and thus were primary candidates for classification as FHB stars. Measures of these stars showed that they had the correct colors and the high radial velocities expected for such stars. Chemical abundance studies were made for two of the stars [HD 161817 - Kodaira (1964) and HD 109995 - Wallerstein and Hunziker (1964)] and the abundances determined were down by a factor of about 10 from the solar abundance, measures in accord with their classification as Pop II objects.

In the Philip (1978) article there is a list of papers concerning the discovery and measurement of FHB stars and measurement of BHB stars up to 1978. Since this time there have been a number of additional papers in which four-color photometry has identified stars which could be members of the FHB class. Additional papers have been identified which proposed FHB candidates on the basis of spectroscopic or other photometric measures. These papers are listed below in Table I.

TABLE I

Additional Papers Listing Candidate FHB Stars

Glaspey, J. W. 1982 *Astrophys. J. Lett.* 258, L71.
Graham, J. A. 1970 *Publ. Astron. Soc. Pacific*, 82, 1305.
Jaschek, M. and Egret, D. 1982 Catalog of Stellar Groups, Pub. Spec. CDS No. 4.
Kilkenny, D. 1984 *Mon. Not. Royal Astron. Soc.* 211, 969.
Philip, A. G. D. 1984 *Contr. of Van Vleck Obs.* No. 2, 1.
Pier, J. 1983 *Astrophys. J. Suppl.* 53, 791.
Sommer-Larsen, J. and Christensen, P. R. 1985 Preprint
Stetson, P. B. 1981 *Astron. J.* 86, 1882.

2. HIGH GALACTIC LATITUDE SURVEYS

I became interested in high galactic latitude blue stars as a result of work started at Warner and Swasey Observatory in the 1960's. A stellar density study was made of a region at galactic latitude -30° and longitude 76° (The 1 HLF 2 region). The interstellar reddening was determined by means of MK spectral types and UBV photometry obtained for the early-type stars found in the area. The derived color excess kept increasing with fainter apparent magnitude until a maximum value was reached, indicating that more distant stars were above the layer of dust in the galactic plane. However there were a number of faint, A stars which had colors too blue for their spectral type and thus when their color indices were corrected for the interstellar reddening the color indices were much bluer than expected (Philip 1972). At about this time the Strömgren four-color system had been calibrated and its use for the study of the nature of early-type stars established. The system was chosen for the study of this group of "too blue" A-type stars. Graham (1970), Philip (1973) and Kilkenny and Hill (1975) published four-color diagrams indicating where stars of various types fell. Stars classified as horizontal-branch stars were distinguished as having c_1 indices 0.20 mag. or more larger than the main sequence values and m_1 indices that were smaller than main sequence values. In the (b-y) vs m_1 plot the indices representing the FHB stars fell along a line orthogonal to the line representing the main sequence A-type stars. The stars that were "too blue" in the 1 HLF 2 region had these same characteristics when measured in the four-color system.

The FHB stars in the 1 HLF 2 region (l = 76° and b = -30°) were found as a result of a complete Schmidt spectral survey, to a limiting magnitude which depended on the prism used at the Schmidt telescope. Each Schmidt plate surveyed a 25 square degree area. All A-type spectra were marked and then measured in the four-color system on later observing runs. By means of the four-color indices the A-type stars could be separated into various groups of Pop I and Pop II stars. Knowing the amount of interstellar reddening in the area, and the absolute magnitudes for each class of star, it was possible to calculate the stellar density distribution for each group.

In a program of galactic structure at high galactic latitude, started at Warner and Swasey Observatory in the 1960's, Schmidt spectral plates have been taken of a number of high galactic latitude areas. In some of these areas the four-color photometry of A-type stars is complete, or nearly complete and many FHB stars have been found. Most of this work was done on AURA telescopes, at CTIO or KPNO. A catalogue of these stars (and some brighter members from the literature) can be found in Philip (1985a). The literature has been searched for four-color measures of other FHB stars (see references in Philip (1978) and Table I in this paper) and plots have been made showing the location of the A-type FHB stars. Over 160 FHB stars have been identified and measured in the four-color system. In Fig. 2 c_1 is plotted versus (b-y) for over 100 FHB stars. The four-color indices have been dereddened according to the estimate of the color excess in each region. In the areas included in the galactic structure survey the reddening was estimated from the four-color and H beta photometry of normal, Pop. I early-type stars. For individual stars taken from the literature, the reddening was estimated from the star's position in

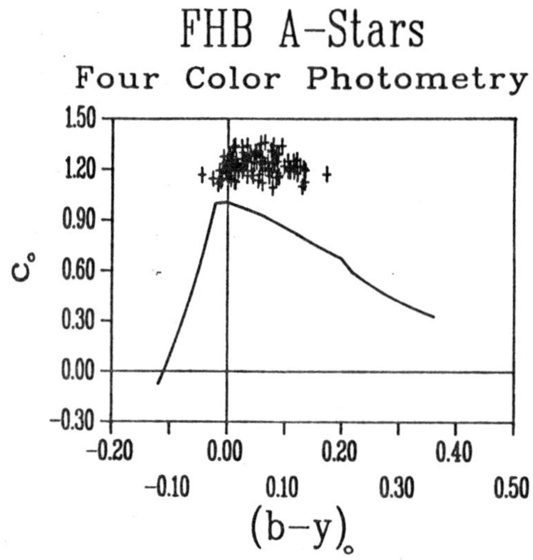

Fig. 2. c_o versus $(b-y)_o$ for FHB Stars.

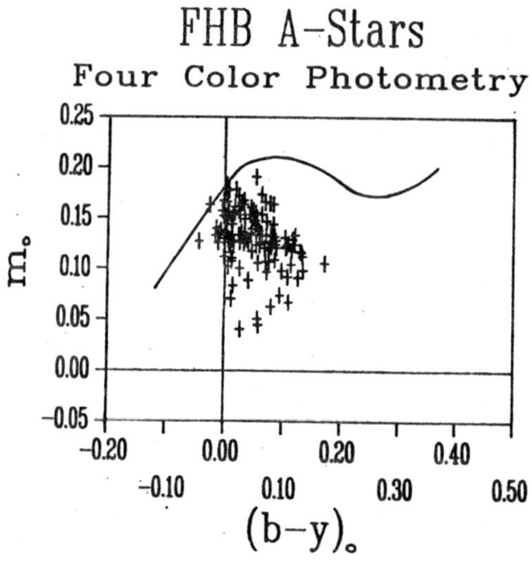

Fig. 3. m_o versus $(b-y)_o$ for FHB Stars.

galactic coordinates. The great majority of these stars is found at high galactic latitudes where the reddening is small, usually under 0.05 magnitudes in (b-y), 0.010 in c_1 and -0.015 in m_1. In Fig 2, a solid line indicates the position of the reference line marking the position of the ZAMS. The indices for individual FHB stars are indicated by crosses. It can be noted that the FHB A-type stars fall in a tight distribution, from 0.2 to 0.4 magnitudes above the reference line.

In Fig. 3, the m_1 index is plotted versus (b-y) and again the indices are dereddened for the effects of interstellar absorption. The symbols are the same as in Fig. 2. The crosses, representing the m_1 indices of FHB stars fall in a band, orthogonal to the relation for normal, Pop. I A-type stars. There seems to be natural cutoff; no indices fall to the right of an inclined line running from upper left to lower right in the diagram. Most of the stars that have indices falling in the lower left of the diagram (at $(m)_1 = 0.04$ and $(b-y)_o = 0.02$) also have c_1 indices higher than 1.35. These stars may not be FHB stars, but rather some type of higher luminosity Pop I star. High dispersion abundance studies could settle the question.

3. THE SIMILARITY OF FHB AND BHB STARS

Using the 90 inch telescope at the University of Arizona, the 60 inch telescope at CTIO and 36 inch telescopes at KPNO and CTIO, a number of BHB stars in the brighter globular clusters have been measured. The best observed clusters in this program are M 4, M 13 and M 55. Figs. 4 and 5 present the c_1 and m_1 indices plotted versus (b-y) for these stars. The colors were dereddened by 0.25, 0.014 and 0.056 mag. in (b-y) for M 4, M 13 and M 55 respectively. It should be noted that the distribution of the indices for the BHB stars in these two diagrams match the distribution of the indices for the FHB stars in Figs. 2 and 3. The c_1 versus m_1 diagrams for FHB and BHB stars are shown in Figs. 6 and 7. Again, the symbols are the same as in Fig. 2. The distributions of the FHB and BHB stars are similar.

Rodgers (1971) reported on radial velocities measured for 60 early-type stars at the South Galactic Pole and Philip (1985b) reported on his velocity measures and measures by Stetson (1981) and Drilling and Pesch (1973) of FHB halo stars. To give an indication of the range in radial velocities found, the radial velocity is plotted versus (b-y) in Fig. 8. The velocity dispersion for the FHB stars in the sample was ±106 km/sec, a unique indicator of Population II. Rodgers (1971) also measured the equivalent width of the calcium K line in the spectra of the SGP A-stars. In Fig 9 the equivalent width of the K line is plotted versus $(b-y)_o$. The solid line represents the relation for Population I early-type stars. The crosses represent the equivalent width of K lines in FHB stars and it can be seen, that as a group, they fall well below the Population I line. This is an effect partly of the lower surface gravity of these high luminosity stars and partly an effect of the lower metal abundance. The abundance effect does not become measurable for stars earlier than spectral type A3.

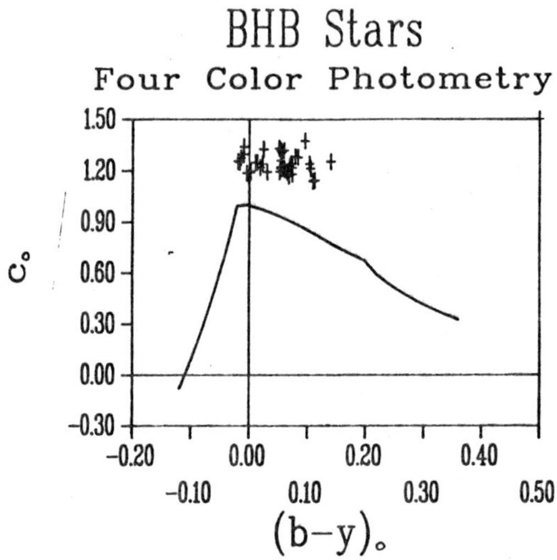

Fig. 4. c_o versus $(b-y)_o$ for BHB Stars.

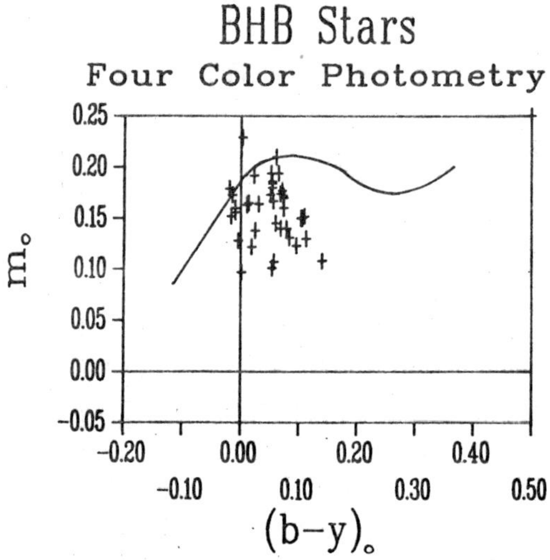

Fig. 5. m_o versus $(b-y)_o$ for BHB Stars.

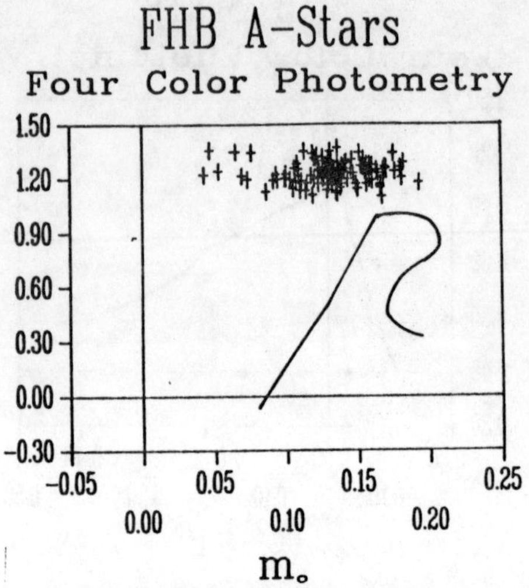

Fig. 6. c_o versus m_o for FHB A-Stars.

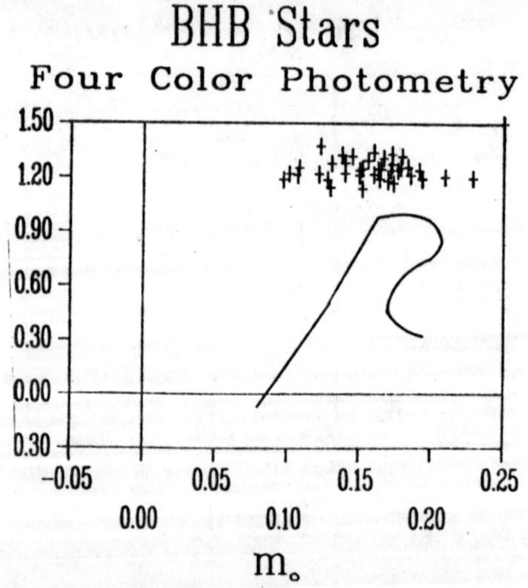

Fig. 7. c_o versus m_o for BHB A-Stars.

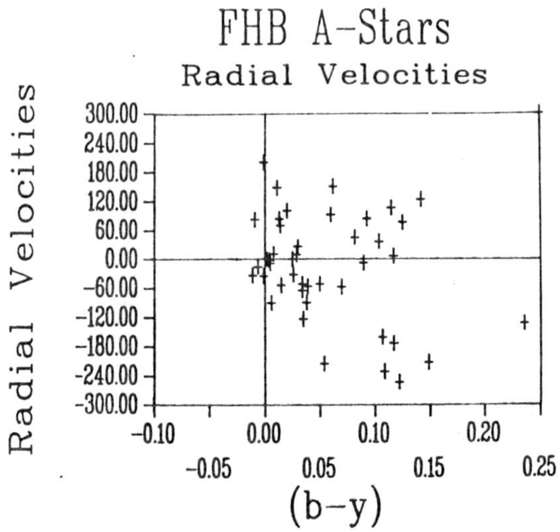

Fig. 8. Radial Velocities of FHB A-Stars.

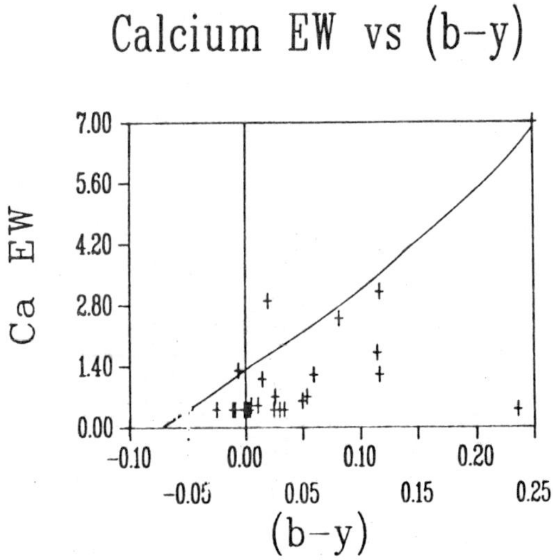

Fig. 9. Calcium K line equivalent width versus (b-y).

Danford (1976) did a spectroscopic study of a number of FHB stars taken from the literature [Roman (1955), Eggen (1964), Newell (1973) and Philip (various lists summarized in Philip (1978))]. He confirmed their nature as FHB stars and obtained abundances of about 1/10 solar (within a range of [Fe/H] of -0.70 to -1.78).

TABLE II

Abundances for FHB A-Type Stars

Star	Abundance	Reference
PS 53 II	-1	Kodaira and Philip (1984)
HD 002857	-1	Kodaira and Philip (1984)
	-1.38	Danford (1976)
HD 014829	-1	Kodaira and Philip (1984)
	-1.78	Danford (1976)
BD 01° 0513	-0.86	Danford (1976)
4 HLF 2a 44	-1.41	Danford (1976)
HD 060778	-1.03	Danford (1976)
HD 074721	-1.11	Danford (1976)
4 HLF 3 16	-0.70	Danford (1976)
HD 086986	-1.11	Danford (1976)
	-1.7	Kodaira (1973)
	-1.43	Kodaira, Greenstein and Oke (1969)
BD 36° 2242	-0.7	Danford (1976)
BD 42° 2309	-1.24	Danford (1976)
HD 109995	-1.9	Kodaira (1973)
	-1.32	Danford (1976)
	-1.24	Kodaira, Greenstein and Oke (1969)
	-1.2	Wallerstein and Hunziker (1964)
HD 130095	-1	Kodaira and Philip (1984)
SS 206 II	-0.61	Danford (1976)
HD 117880	-1.73	Danford (1976)
1 HLF 5 10	-0.77	Danford (1976)
HD 161817	-1	Kodaira and Philip (1984)
	-1.20	Danford (1976)
	-1.6	Kodaira (1973)
	-1.28	Wallerstein and Hunziker (1964)
	-1.21	Kodaira (1964)
	-0.98	Wallerstein, Stone and Williams (1962)
	-0.54	Aller and Greenstein (1960)
HD 213468	-0.84	Danford (1976)

Positions and finding charts for all non-HD or BD stars in the list above can be found in Philip (1985a)

Kodaira and Philip (1984) reported on a high dispersion study of FHB stars. For five FHB stars they found [Fe/H] = -1. Various estimates have been made for the metal abundance of the FHB stars HD 86986, 109995 and 161817. [Fe/H] values are summarized in Table II, shown on the previous page.

Spectrophotometric scans were made of BHB and FHB stars by Hayes and Philip (1983) and sets of FHB and BHB stars were found with nearly identical energy distributions between 3500 and 6800 Angstroms. These scans are shown in Fig. 10. which is taken from Fig. 2 of the Hayes and Philip paper. The BHB star III 17 in M 5 is similar to the FHB

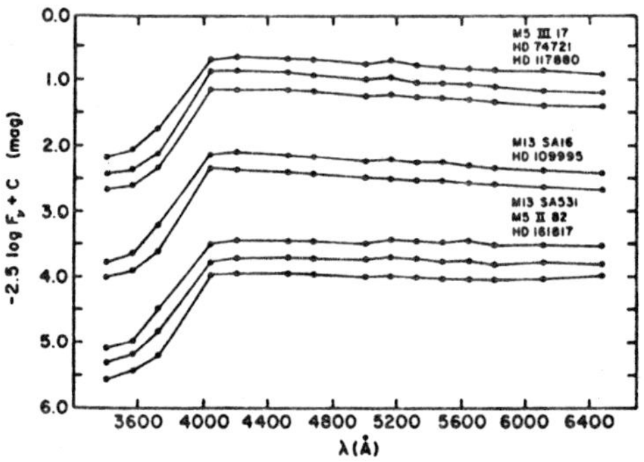

Fig. 10. Intensity Distributions for FHB and BHB stars.

stars HD 74721 and 117880, the star SA 16 in M 13 is similar to HD 109995 and HD 161817 is similar to the BHB stars SA 531 in M 13 and II 82 in M 5.

For the reasons given in the paragraphs above, the evidence supports the idea that the FHB stars are quite similar to the BHB stars in globular clusters. In so far as this idea is correct, then one can investigate the brighter FHB stars at high dispersion and in effect study the nature of the fainter BHB stars. So far no FHB stars have been found with [Fe/H] < -1.7 although there are many globular clusters with [Fe/H] < -2.0 or - 2.2. Whether this is a real effect or just the result of a small sample of FHB stars with measured abundances is yet to be determined.

4. EXTENDING THE SURVEY

In December, 1984, an observing run was made on the Michigan Curtis Schmidt Telescope at CTIO. The thin prism was used and unwidened spectra were taken with exposures of two hours each. In this way spectra were obtained to a visual magnitude of V = 16 to 17 at the South Galactic Pole and in 4 HLF 6, a region at l = 180° and b = -60°. For the brightest stars, A stars could be identified by the pinching effect for each of the hydrogen Balmer lines. But for the rest of the stars no lines could be seen. The new plates were taken in regions already surveyed with the thin prism, with widened spectra. In the magnitude regions that overlapped, it was possible to identify A-type stars and see how they appeared on the unwidened plates. A-type stars had spectra with flat continua, except for a slight dip caused by the sensitivity function of the IIIa-J emulsion. From 100 - 200 A-type stars were found per plate, which is a great increase in the numbers of A-type stars found on the older plates, which were complete to a visual magnitude of near V = 14.

In Philip (1972) there was a figure in which the stellar density distributions of Population I A-type stars, Population II A-type stars and RR Lyrae stars (Kinman et al. 1966) at the North Galactic Pole were plotted. The plot showed the distribution up to 4 kpc from the galactic plane. In Fig. 11 a new stellar density distribution is shown including the new stars found on thin prism plates. Bruce Twarog

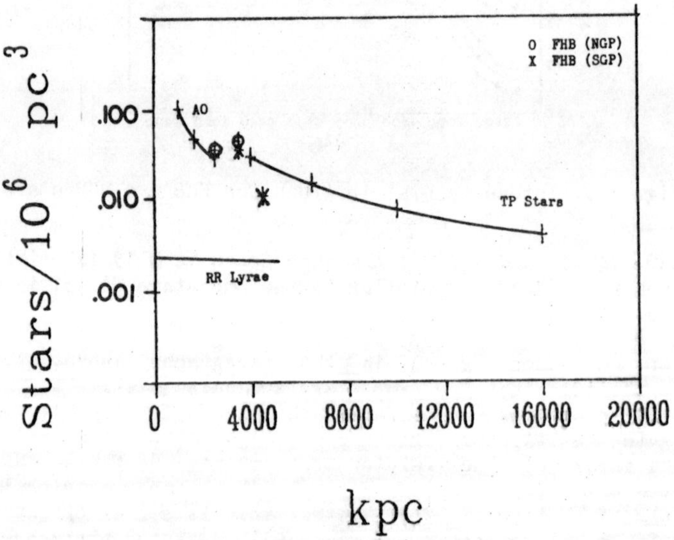

Fig. 11. Stellar Density Distribution for early-type stars in the galactic halo.

measured one of the new A-type stars in the four-color system on the CTIO 4 m telescope. He found colors (b-y) = 0.11, c_1 = 1.5 and m_1 = 0.17 and a V magnitude of 15.6. These colors do not agree with the classification of these stars as Pop I main sequence stars or as subdwarf A stars. The stellar density plot was made on the assumption that the newly identified faint A-type stars are Pop II, halo stars. Pier (1983) found that most of the AB type stars found in the Preston and Schectman (1980) survey, in the magnitude range 13.5 - 15.5 were FHB stars, after the Am, Pop I and high luminosity stars were culled from the sample. Pier found no Pop I stars further than 3 kpc from the galactic plane. Thus it seems to be a reasonable assumption that the majority of the faint A-type stars between V magnitudes of 14 - 17 are not Pop I stars, but Pop II stars. Additional photometric observations are planned of a sample of the stars found on the unwidened spectral plates in order to confirm their population type. Since magnitudes have not yet been determined for these stars, they were apportioned in the magnitude intervals 14-15, 15-16 and 16-17 according to the way in which FHB stars were apportioned in the three earlier magnitude intervals. Based on these assumptions, the A-type FHB stars have a density distribution in the galactic halo much greater than that of the field RR Lyrae stars, which is what one finds in those globular clusters that have RR Lyrae stars.

5. CONCLUDING REMARKS

Studies of globular clusters can provide data for testing theories of stellar evolution. The predictions of evolutionary theory, applied to a group of stars formed at a given time with a given chemical composition, can be compared to the distribution of points in numerous diagrams obtained from observations, such as color-magnitude diagrams or log L, log T_{eff} diagrams. At present there is much activity in fitting theoretical main sequences with observed main sequences in order to obtain age and abundance estimates for globular clusters.

High dispersion spectra can yield evidence concerning the abundances of various elements. There exist data showing that, in some clusters, there are abundance variations within the cluster. An important question to be answered is whether these abundance variations are primordial or if they are the result of mixing at some stage during stellar evolution. It may be that combinations of both effects exist. Another important question is that of the second parameter affecting the morphology of CM diagrams. There exist clusters with the same [Fe/H] values which have different morphologies in the HR Diagram. Many parameters have been suggested (helium, CNO and age, for example) but a definite explanation is not yet at hand.

Since the globular clusters are all more than 2 kpc from us, it is difficult to obtain data at a high enough resolution. If the FHB stars are the same population as the BHB stars, then they are very important, for these stars are quite bright (up to V = 7.0 mag.) and very high dispersion work can be done on them. The photometric data, summarized in section 3, seem to indicate many similarities between the two groups of stars. Kraft (1978, 79) reviewed spectroscopic data for field and cluster giants. There was no strong evidence for differences in [X/H] for most elements but he quoted Sneden's (1974) result that the average nitrogen abundance in M 92 giants is considerably higher than that for field, halo giants. Kraft suggests that the halo field stars could have come from the present globular clusters or they may have come from clusters that have broken up completely. These clusters could have had a different enrichment history than that of the present globular clusters.

The position of a HB star in the HR diagram is dependent on many factors, among which are (in addition to the ones mentioned above for the second parameter problem) rotation, core mass, mass loss and CNO abundance. What is needed is more, high quality data concerning the individual parameters for Pop II stars so that limits can be placed on the range of possible variations of the parameters. Buonanno et al. (1985) have pointed out that one way to proceed is to consider simultaneously various parameters. One might study a group of clusters in which some of the basic parameters are assumed to be the same.

Until firm data become available to show that the field stars are different from the cluster horizontal-branch stars, a way to proceed is to obtain high dispersion data for the brighter FHB stars, in the UV, visible and IR, as a way to obtain observational data on the abundances of many elements in HB stellar atmospheres. These derived abundances will serve as good tests of stellar evolution theory.

REFERENCES

Aller, L. M. and Greenstein, J. L. 1960 Astrophys. J. Suppl. 5, 139.
Arp, H. C. 1962 Astrophys. J. 135, 311.
Buonanno, R., Corsi, C. E. and Fusi Pecci, F. 1985 Astron. Astrophys. 145, 97.
Danford, S. 1976, Thesis, Yale University.
Drilling, J. S. and Pesch, P. 1973 Astron. J. 78, 47.
Eggen, O. J. 1964 Royal Obs. Bull, 84.
Eggen, O. J. 1972 Astrophys. J. 172, 639.
Feige, J. 1958 Astrophys. J. 128, 267.
Hayes, D. S. and Philip, A. G. D. 1983 Astrophys. J. Suppl. 53, 759.
Humason, M. L. and Zwicky, F. 1947 Astrophys. J. 105, 85.
Kilkenny, D. and Hill, P. W. 1975 Monthly Notices Roy. Astron. Soc. 173, 625.
Kinman, T. D., Wirtanen, C. A. and Janes, K. A. 1966 Astrophys. J. Suppl. 13, 379.
Kodaira, K. 1964 Ztschr. Astrophys. 59, 139.
Kodaira, K. 1973 Astron. Astrophys. 22, 273
Kodaira, K., Greenstein, J. L. and Oke, J. B. 1969 Astrophys. J. 155, 525.
Kodaira, K. and Philip, A. G. D. 1984 Astrophys. J. 278, 208.
Kraft, R. P. 1978 in Globular Clusters, D. Hanes and B. Madore, eds., Cambridge Press, Cambridge, p. 87.
Kraft, R. P. 1979 in Annual Review of Astron. Astrophys. 17, 309.
Newell, B. 1973 Astrophys. J. Suppl. 228, 37.

Oke, J. B., Greenstein, J. L. and Gunn, J. 1966 in Stellar Evolution, R. F. Stein and A. G. W. Cameron, eds., Plenum Press, New York, p. 399.
Philip, A. G. D. 1972 in Stellar Evolution, H. Y. Chiu and A. Muriel, eds., MIT Press, Cambridge, p. 397.
Philip, A. G. D. 1972 in The Role of Schmidt Telescopes in Astronomy, U. Haug, ed., p. 117.
Philip, A. G. D. 1973 in IAU Symposium No. 50, Spectral Classification and Multicolor Photometry, Ch. Fehrenbach and B. Westerlund, eds., Reidel, Dordrecht, p. 230.
Philip, A. G. D. 1978 in IAU Symposium No. 80, The HR Diagram, A. G. D. Philip and D. S. Hayes, eds., Reidel, Dordrecht, p. 209.
Philip, A. G. D. 1985a Van Vleck Observ. Contr. No. 2, 1.
Philip, A. G. D. 1985b in IAU Colloquium No. 88, Stellar Radial Velocities, A. G. D. Philip, ed., L. Davis Press, Schenectady, p. 359.
Philip, A. G. D., Cullen, M. F. and White, R. E. 1976 UBV Color-Magnitude Diagrams for Galactic Globular Clusters, Dudley Obs. Report No. 11.
Rodgers, A. W. 1971 Astrophys. J. 165, 581.
Roman, N. 1955 Astrophys. J. Suppl. 2, 195.
Simoda, M. and Tanikawa, K. 1970 Pub. Astron. Soc. Japan, 22, 143.
Sneden, C. 1974 Astrophys. J. 189, 493.
Stetson, P. B. 1981 Astron. J. 75, 246.
Wallerstein, G. and Hunziker, W. 1964 Astrophys. J. 140, 214.
Wallerstein, G., Stone, Y. H. and Williams, J. A. 1962 Astrophys. J. 135, 459.

DISCUSSION

JANES: Do you have some cut off in the (b-y), c_1 diagram?

PHILIP: The initial cut is from the spectra. The candidate star has to be of A-type. Then in the (b-y), c_1 diagram the star should have (b-y) colors between 0.00 and about 0.18.

JANES: I was wondering about the cutoff in the other dimension.

PHILIP: My definition is that delta c_1 has to be about 0.2 or greater.

RENZINI: The lower boundary is a real feature?

PHILIP: I am not willing to say this for the FHB stars. It looks as if it is in the diagram but you must remember that the stars selected for this plot must have delta c_1 values greater than 0.2. In the BHB star plot the lower envelope is real because here the stars were selected because they were identified as BHB stars from a UBV color-magnitude diagram and then measured in the four-color system. Thus here there is no selection according to delta c_1.

RENZINI: One expects a gap between Pop II field A stars or BHB stars and the Pop I field stars in that type of diagram. In the metallicity range between -1.3 or so and 0.0 you don't have blue horizontal-branch stars at all.

SPECTRA OF THREE PROTOTYPE FIELD HORIZONTAL-BRANCH A STARS

A. G. Davis Philip*

Van Vleck Observatory and Union College

John Lee

Van Vleck Observatory

ABSTRACT. Spectra, at 9 Ångstroms/mm, obtained at the Soviet Union's Six Meter reflector, are presented for the FHB prototype stars HD 86986, 109995 and 161817. Portions of tracings from λ 4302 to 4550 are shown.

1. INTRODUCTION

Spectra, at 9 Ångstroms/mm, have been obtained at the main spectrograph of the Soviet Union's 6 meter telescope. The spectra were obtained by Philip and N. Samus during an observing run in August, 1980 and by V. Panchuk of the SAO staff in March and April of 1981. The plates were traced on the PDS microdensitometer at Kitt Peak National Observatory by Philip. A stellar atlas of field horizontal-branch stars is in preparation. Photographic copies of the spectra of HD 86986, 109995 and 161817 will be presented along with co-added tracings of each spectrum. Equivalent widths have been calculated for lines with equivalent widths greater than 30 milliangstroms. In this paper we present copies of the photographic spectra and sample regions of the spectral tracings.

2. FHB STARS

As is explained in the previous article (Philip 1985), the field horizontal-branch stars have many of the same characteristics as do blue horizontal-branch stars in globular clusters. These similar characteristics are: the four-color indices of the FHB stars are the same as those for BHB stars, spectrophotometric measurements of some FHB and BHB stars are within 0.02 magnitudes of each other over wavelength

---

* Visiting Astronomer, Soviet Academy of Sciences, Special Astrophysical Observatory and Kitt Peak National Observatory.

A. G. Davis Philip, (ed.)
Horizontal-Branch and UV-Bright Stars 57–62

intervals in the region of 3500 to 6800 Ångstroms, the velocity dispersion of a group of FHB stars is over ±110 km/sec, the calcium "K" line equivalent widths fall well below the main sequence values and [Fe/H] values obtained so far for FHB stars fall in the range of -1 to -2. For all of these reasons it seems that FHB stars are the same type of star as the BHB stars and thus by studying the brighter FHB stars one can obtain much important information concerning the much fainter BHB stars.

It was for this reason that the plan was made to obtain nine Ångstrom/mm spectra of the prototype FHB stars and to prepare a spectral atlas as a guide for those studying horizontal-branch A-type stars in globular clusters and in the field. The three prototype FHB stars are the ones analysed by Oke Greenstein and Gunn (1966). They obtained photoelectric scans from about 3300 to 10000 Ångstroms which were matched to atmospheric models to obtain temperatures and gravities. Among the stars which matched the characteristics of horizontal-branch stars were the three listed in the first paragraph. In the years since much additional work has been done on these stars, including high dispersion spectra for abundance analyses, four-color photometry, UBV photometry and new photometric scans. These three stars are now among the best observed stars, of Population I or II.

3. SPECTRA AND TRACINGS

Photographic reproductions of the spectra of HD 109995, 86986 and 161817 (in order of decreasing temperature) are displayed in Fig. 1. Some of the stronger lines are identified at the bottom of the figure. After the hydrogen lines and the K line of calcium II, the next strongest lines, due to a single element (ranked in order of its strength in HD 161817), are those of Fe II (λ4549.50), Mg II (λ4481.23), Fe I (λ4383.55 and 4045.82), Ca I (λ 4226.00) and Ti II (λ 4395.03, 4468.49 and 4300.05). For each of these lines, the line is strongest in HD 161817 and weakest in HD 109995 (primarily an effect of temperature).

Portions of the tracings of the three spectra are shown in Figs. 2a and 2b. H gamma appears on the left side of the tracing in Fig. 2a and a line of Ti II at λ 4443.8 appears on the right. In Fig. 2b the line of Ti II at λ4443.8 appears at the left and a blend of two Fe II lines at λ4549.5 appears on the right. In this spectral region lines of FE I and II, Ti II, Ca I and Sc II appear as unblended lines and Mg II as a blended line. These lines are strongest in the spectrum of HD 161817 (cooler) and weakest in the spectrum of HD 109995 (hotter).

The full spectral atlas is being readied for publication and will be published at a later time.

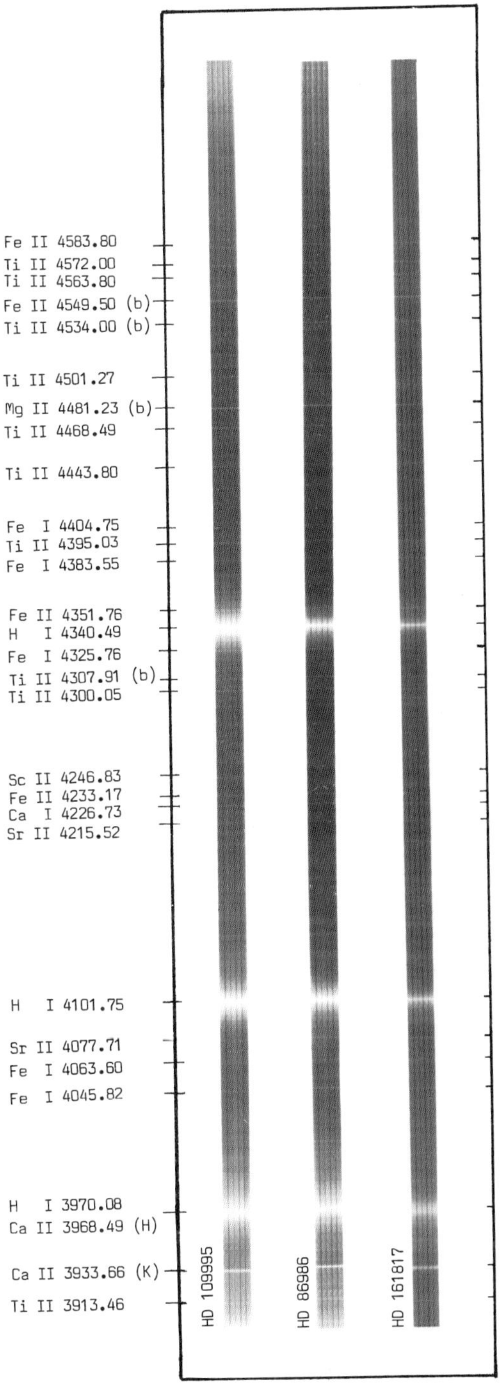
Fe II 4583.80
Ti II 4572.00
Ti II 4563.80
Fe II 4549.50 (b)
Ti II 4534.00 (b)
Ti II 4501.27
Mg II 4481.23 (b)
Ti II 4468.49
Ti II 4443.80
Fe I 4404.75
Ti II 4395.03
Fe I 4383.55
Fe II 4351.76
H I 4340.49
Fe I 4325.76
Ti II 4307.91 (b)
Ti II 4300.05
Sc II 4246.83
Fe II 4233.17
Ca I 4226.73
Sr II 4215.52
H I 4101.75
Sr II 4077.71
Fe I 4063.60
Fe I 4045.82
H I 3970.08
Ca II 3968.49 (H)
Ca II 3933.66 (K)
Ti II 3913.46
HD 109995
HD 86986
HD 161817

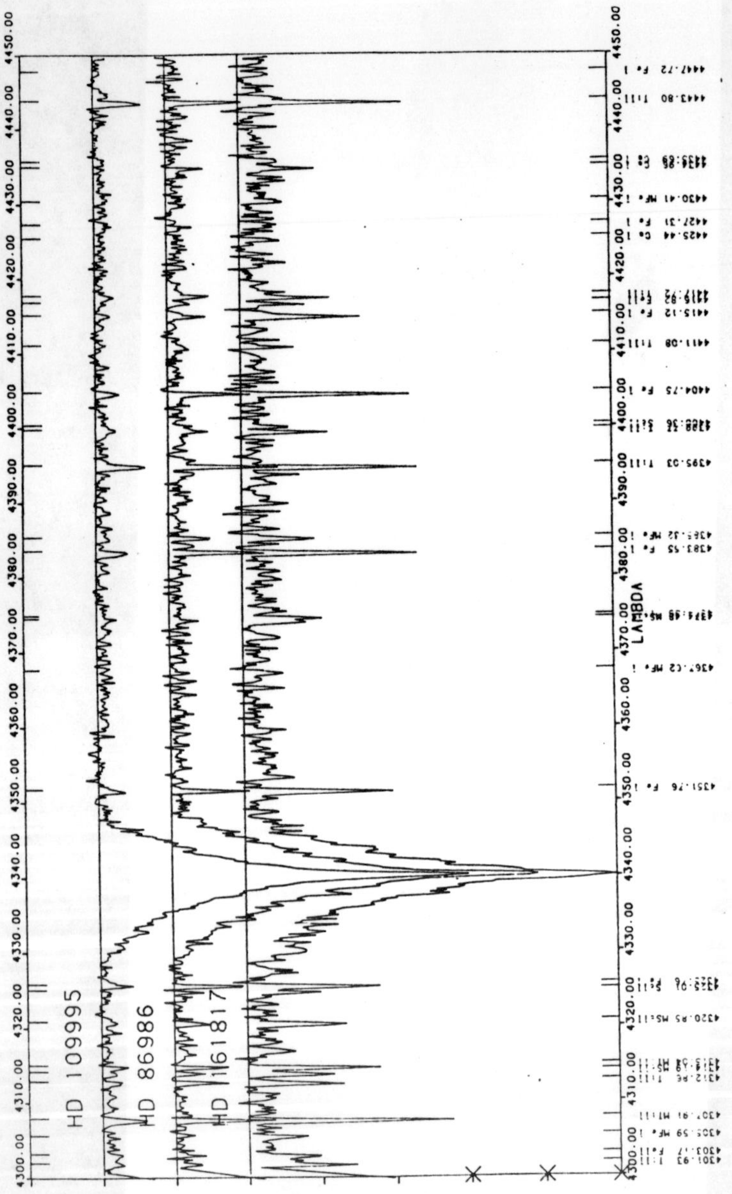

Fig 2a. Tracing of spectra of the three stars in Fig. 1. from λ 4302 to 4444.

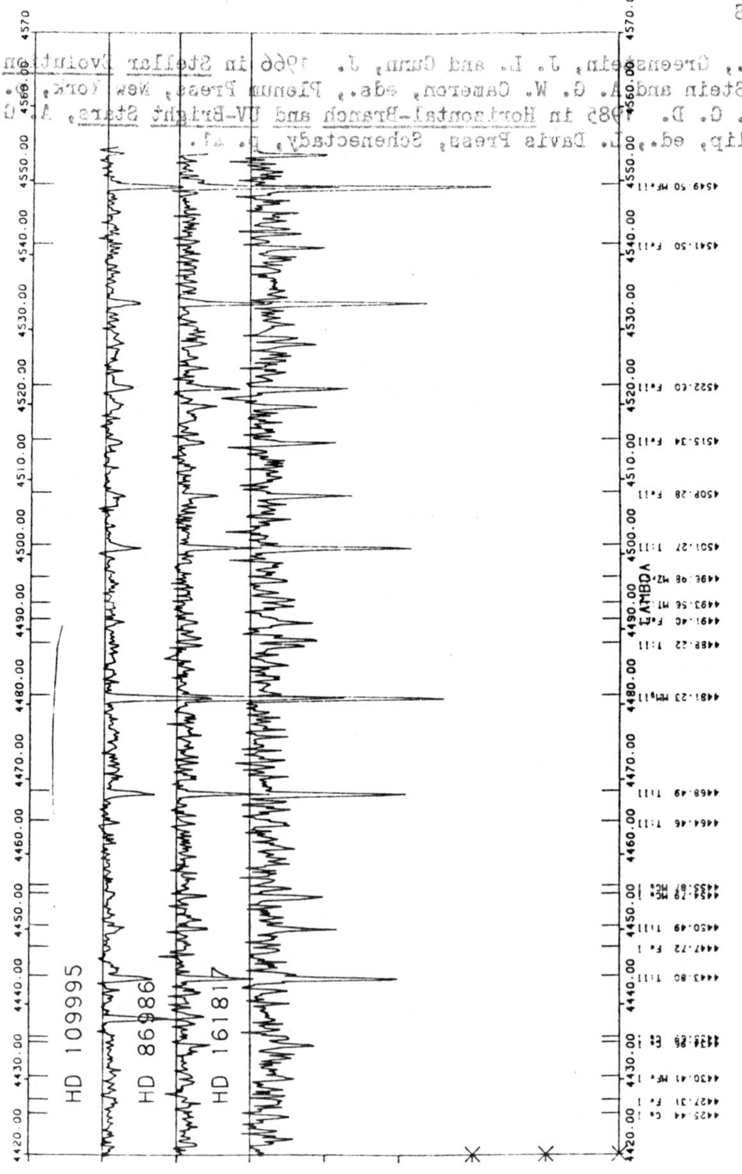

Fig. 2b Tracing of spectra of the three stars in Fig. 1. from λ 4425 to 4550.

REFERENCES

Oke, J. B., Greenstein, J. L. and Gunn, J. 1966 in Stellar Evolution, R. F. Stein and A. G. W. Cameron, eds., Plenum Press, New York, p. 399.
Philip, A. G. D. 1985 in Horizontal-Branch and UV-Bright Stars, A. G. D. Philip, ed., L. Davis Press, Schenectady, p. 41.

HORIZONTAL-BRANCH A-STARS: TEMPERATURES AND SURFACE GRAVITIES

D. S. Hayes

Kitt Peak National Observatory
National Optical Astronomy Observatories*

ABSTRACT. This paper summarizes recent work on the effective temperatures and surface gravities of horizontal-branch A-stars. The effective temperatures and surface gravities are obtained by fitting observed absolute energy distributions to energy distributions calculated using ATLAS model atmospheres. With the additional knowledge of the absolute bolometric magnitude, the mass of each star may be calculated. The resulting values are compared with masses of double-mode RR Lyrae stars, and with evolutionary tracks, and this comparison is used to evaluate the accuracy of the surface gravities.

1. INTRODUCTION

The temperatures, gravities and masses of horizontal-branch (HB) stars have been discussed previously (Hayes and Philip 1979), using four-color photometry by Philip (1983) and energy distributions for seven stars found in the literature. In the beginning, we had two objectives for this work. One was to calibrate the four-color system in terms of T_{eff}'s and log g's, so that statistical surveys of these stars in the theoretical HR diagram could be made. The other was to exploit the T_{eff}'s and log g's to derive semi-empirical masses so that tests of evolutionary theory could be made. In order to calculate masses, one must have the absolute bolometric magnitude; the absolute visual magnitude is derived from the morphology of globular cluster horizontal branches plus the absolute visual magnitudes of the RR Lyrae stars and the bolometric correction is derived from model atmospheres (see Hayes and Philip 1979). The T_{eff}'s and log g's are obtained by fitting the energy distributions to ATLAS model atmospheres with metallicities one-tenth solar (Kurucz 1979, 1981). The masses are then calculated: $M = gL/T^4_{eff}$. Thus, the masses we obtain are <u>semi</u>-empirical; they depend on the model atmospheres in a critical way.

Hayes and Philip (1979) found derived masses for seven stars, and obtained a mean mass of $0.6\ M_\odot \pm 0.4\ M_\odot$. When the observations are compared with models of the ZAHB and with post-ZAHB evolutionary tracks

---

* Operated by AURA under contract with the National Science Foundation

A. G. Davis Philip, (ed.)
Horizontal-Branch and UV-Bright Stars 63–74

by Sweigart and Gross (1976), there is good agreement in the sense that, in the (log g, log T_{eff}) and (log L, log T_{eff}) diagrams, the mean values correspond to the locations of theoretical HB stars with masses of about 0.6 $M_\odot$, and with reasonable values of helium abundance and metallicity. This is also true for the data discussed in this paper. But the range of derived mass is quite large, whereas the ZAHB models and post-ZAHB evolutionary tracks show that, for the range in T_{eff} being dealt with here, the range in mass should be quite small (about ±0.02 $M_\odot$ at most). Furthermore, the derived masses show a trend, in that the coolest stars (T_{eff} about 7500 K) show very low masses (about 0.24 $M_\odot$), whereas the hottest stars (T_{eff} about 8600 K) show very large masses (about 1.08 $M_\odot$). This strong trend in derived mass is in disagreement with the evolutionary models. Of course, in a group of real field HB stars one would expect a range of masses to be present as a result of a range of ages and initial parameters such as helium abundance, metallicity, angular momentum and so forth. But one would not expect as large a range of mass as is shown by the derived masses, and there is no reason to expect the correlation of mass with T_{eff} which is found. Hayes and Philip (1979) concluded that this "discrepancy" was a result of some deficiency in the physics of the model atmospheres used, so that the values of log g which were obtained were erroneous.

Meanwhile, scanner energy distributions of much higher quality for a much larger sample of stars have been obtained. In this paper the procedure developed by Hayes and Philip (1979) is applied to these data, and the results are discussed.

2. THE OBSERVATIONS

The energy distributions of 23 horizontal-branch (HB) A-stars, including six globular cluster stars and 17 field stars have been measured. Twelve field stars, ranging from 7.0 to 10.8 magnitude, were observed with the HCO scanners on 0.9-meter telescopes at KPNO and CTIO (Philip and Hayes 1983). Seven field stars, ranging from 7.0 to 12.4 magnitude, (including two observed at KPNO and CTIO) and six globular cluster stars, ranging from 14.9 to 15.5 magnitude, were observed with the Oke multichannel spectrometer on the 5-meter telescope at Palomar Mountain (Hayes and Philip 1983).

The observations were made at wavelengths between 3400 and 6800 Å. The National Optical Astronomy Observatories (NOAO) measurements were referred to the Breger (1976a) standards ξ^2 Cet, η Hya and 109 Vir. the Palomar measurements were referred to Oke's multichannel spectrophotometric standard BD +17°4708 (Oke and Gunn 1983). There were four stars common to the NOAO and Palomar programs; the agreement is excellent (±0.016 mag., rms.) and shows that there is no significant difference in the absolute calibrations. All observations are referred to the absolute calibration of Hayes and Latham (1975). Atmospheric extinction was carefully measured.

One question which we must deal with is whether the FHB A-stars are really the same kind of stars as those found in globular clusters. Philip (1978, 1985) has shown that the two groups are similar in the four-color system. On the basis of the scanner data, which contains more information that four-color photometry, Hayes and Philip (1983) and Philip (1985) show that the field and cluster stars are indistinguishable spectrophotometrically.

3. TEMPERATURES AND SURFACE GRAVITIES

The process of deriving T_{eff} and log g from the scanner energy distributions by fitting to model atmospheres is as follows: values of the colors (3636 - 4036) and (4036 - 6790) are derived from the scans; they are dereddened, and then they are plotted in a diagram in which the corresponding colors from a grid of fully-blanketed ATLAS model atmospheres are also plotted. Graphical interpolation between the values of T_{eff} and log g for the models is used to derive the values for the stars. The models used in this investigation are ATLAS models which are similar to those published (Kurucz 1979), but with an improved treatment of convection (Kurucz 1981, Lester et al. 1982).

This process of deriving T_{eff} and log g from the two colors (3636 - 4036) and (4036 - 6790) does not use all the information in the scans, but, in view of the discrepancy which was bound by Hayes and Philip (1979), it will be used until more appropriate model atmospheres make a more sophisticated fitting procedure worthwhile.

As a check of the validity of this process of deriving values of T_{eff} and log g from ATLAS model atmospheres, it has been applied to Population I stars. The correctness of the resulting values of log g are of primary concern, since it appears that the discrepancy for the HB A-stars is primarily in the values of log g obtained. Breger's (1976b) catalog of scanner energy distributions was searched for main-sequence (class V) stars in the range of spectral types from late B through early F.

Values of the colors (3636 - 4036) and (4036 - 5556) were derived for 58 stars; they are shown plotted in Figure 1. Note that the color (4036 - 5556) is used instead of (4036 - 6790), since Breger's catalog does not extend to the longer wavelength for a large enough sample of stars. Also shown in Figure 1 is the ATLAS model grid for solar abundance, along with the reddening line for $E_{(B-V)} = 0.10$ mag. No dereddening corrections have been applied to the scanner data.

Although the scatter is fairly large, it is clear that the stars follow a line of constant gravity well, and that a value of log g ~ 4.0 would be a good fit for the mean line. Two of the stars are obviously reddened, but the remainder do not show significant reddening. This is not surprising, since most of these stars are relatively bright. In particular, the relatively large scatter around $T_{eff} \backsim 9500$ K (A0 V stars) is not plausibly explained by reddening.

Forty-five of these stars, with -0.06 < (B-V) < +0.28, have been selected and values of log g have been derived by graphical interpolation in the model grid. The mean value of log g obtained in this way is 3.93 ± 0.24. These results can be compared with values of log g obtained from eclipsing binaries. Popper (1980) lists values of mass and radius for well-observed detached main-sequence eclipsing binaries. Log g was calculated for 33 stars in 18 systems ranging from late B through early F types. The stars were selected to have $(B-V)_o$ between +0.105 and +0.40 mag. When plotted, three stars showed significantly lower values of log g than the remainder; they were eliminated on the basis that they were probably somewhat evolved. The remaining 30 stars have a value of log g = 4.14 ± 0.15 $M_\odot$, with a slight trend towards lower values for the cooler stars. In Figure 1 the dashed line shows roughly the locus of models with log g = 4.14. Clearly, although the scanner observations correspond better with log g = 3.93, the difference is not large compared to the scatter in the observations, and there is no evidence of a significant trend with T_{eff}. This difference of 0.2 in log g corresponds to a shift of about 0.045 mag. in the color (3636 - 4036) at a T_{eff} of about 8750 K, in the sense that the models have the smaller Balmer discontinuities.

The accuracy of the effective temperatures from the grid can be tested in the same way. For the same 45 Population I stars, values of T_{eff} have been obtained by graphical interpolation in the model grid. The comparison values of T_{eff} have been found using the fundamental effective-temperature scale constructed by Hayes (1978). That scale is given in terms of (B-V), so the values of (B-V) given by Breger in his catalog have been used to interpolate an effective temperature for each star. Then the difference ($T_{eff}^{SCANS} - T_{eff}^{FUND}$) is calculated. The mean value is 0 ± 185 K. There is no apparent trend of the difference with (B-V). Thus, for Population I stars, using the colors derived from scans to interpolate in the ATLAS model grid produces excellent results.

The difference in Balmer discontinuity of 0.045 mag., found above, is not large, in view of the scatter in the observations, but is worth discussing. There are two likely sources of this shift, plus some minor contributors. The two likely sources include a problem in the predicted spectrum of the models, and a possible systematic error in the absolute calibration of Vega. First, with respect to the models, Hayes (1985) shows a fit of one of the Kurucz (1979) models (9400, 3.95, 1.0) to a new mean absolute energy distribution of Vega. The model fits the observations well, except in the wavelength interval from 4000 to 5000 Å. With the model and observations in agreement at λ < 3700 Å and λ > 5000 Å (extending to 7500 Å), the model is fainter than the observations by about +0.05 mag. at 4000 Å, with the difference decreasing with increasing wavelength to about 5000 Å, where they agree. In terms of the color (3636 - 4036), the model has a smaller Balmer discontinuity by about 0.04 mag. This is in very good agreement with the shift discussed above. On the other hand, the <u>systematic</u> accuracy of the absolute calibration of Vega is on this same order of size. Hayes and Latham

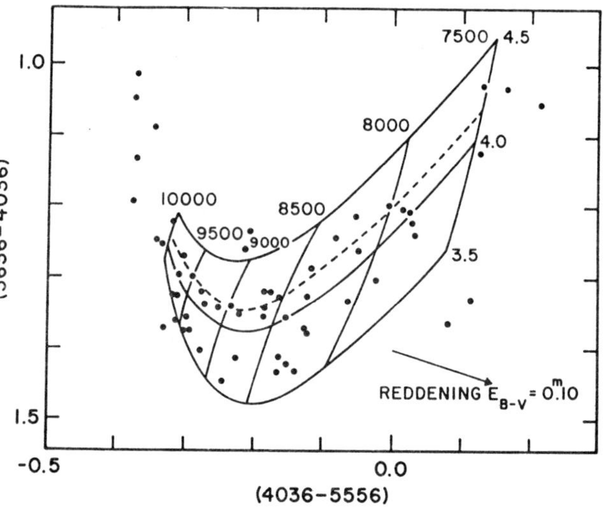

Fig. 1. A color-color diagram with (3636 - 4036) plotted against (4036 - 5556). The points are main-sequence stars from the catalog of scanner energy distributions by Breger (1976b). The grid is from solar-abundance ATLAS model atmospheres.

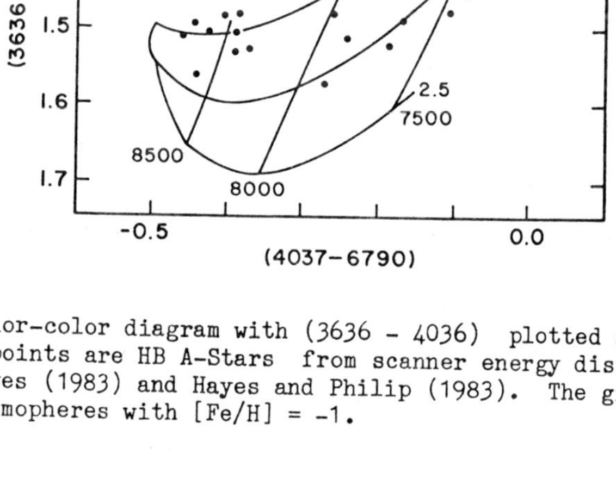

Fig. 2. A color-color diagram with (3636 - 4036) plotted against (4036 - 6790). The points are HB A-Stars from scanner energy distributions by Philip and Hayes (1983) and Hayes and Philip (1983). The grid is from ATLAS model atmopheres with [Fe/H] = -1.

(1975) estimate the accuracy of the color (3636 - 4036) in the absolute calibration of Vega (to which Breger's and our own data are referred) to be ±0.03 mag. Additional systematic errors can arise from the calibration of the secondary standards relative to Vega, and in the spectrophotometry of the program stars. Other contributing factors to the shift could be the rather loose criteria for selection of the stars (the sample in Breger's scanner catalog is not homogeneous in terms of observational methods, metallicity, luminosity classification, etc.), mismatches in the passband placement and width in the observations as opposed to the model, and so forth. Clearly, each of these effects could be explored further. The disagreement between the energy distribution of the model and the new absolute calibration of Vega found by Hayes (1985) suggests that there are minor problems with the model atmosphere. Nevertheless, we can conclude that the shift we have found here, and the possible causes of the shift, do not explain the discrepancy we find in Population II stars. The method we are using appears to be sound, and the systematic accuracy (determined by the absolute calibration of Vega and the transfers of the standard stars, etc.) is about 0.2 in log g.

In Figure 2 the color-color diagram is shown for the colors (3636 - 4036) and (4036 - 6790) for the HB A-stars. Also shown is the grid of ATLAS models for [Fe/H] = -1, and the reddening line. The observed points have been dereddened. The details of the dereddening process, and a table of the values obtained, may be found in Hayes and Philip (1986). From the positions of the points in Figure 2 values of T_{eff} and log g have been interpolated; and these are shown plotted for each star against $(b-y)_o$ in Figures 3 and 4. In Figure 3, in which T_{eff} is plotted, two intersecting straight lines have been fitted to the data. These lines are given by:

$$T_{eff} = -16143(b-y)_o + 9030° \text{ K}$$

$$\text{where } 0.01 \leq (b-y)_o \leq 0.07 \text{ mag.,}$$

and

$$T_{eff} = -9857(b-y)_o + 8590° \text{ K,}$$

$$\text{where } 0.07 \leq (b-y)_o \leq 0.14 \text{ mag.}$$

The points scatter about this double-segment line with a standard deviation of ± 144° K. Since the temperature can be expected to be relatively unaffected by whatever is the source of the discrepancy in log g, the temperature scale given by these lines can be recommended for use in such work as abundance studies from high-dispersion spectra.

The surface gravities are shown plotted in Figure 4. A "best fit" (eyeball) line is shown, along with two dashed lines giving the locus of stars of masses 0.60 $M_\odot$ and 1.00 $M_\odot$, respectively. In the case of the

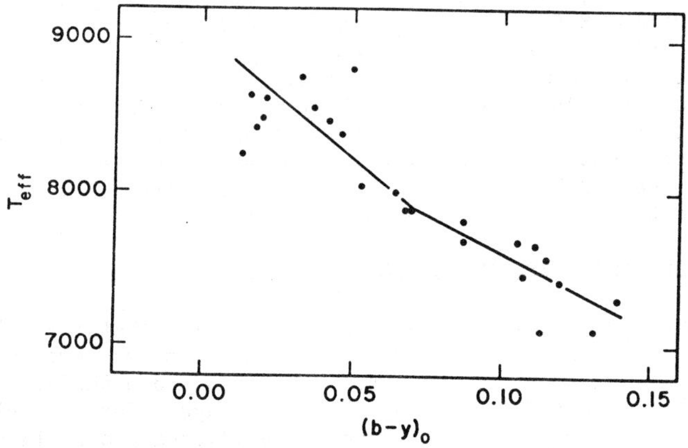

Fig. 3. T_{eff} is plotted against $(b-y)_{0}$ for HB A-stars; the values are derived from Fig. 2. The lines are "eyeball" best-fit lines; the equations are given in the text.

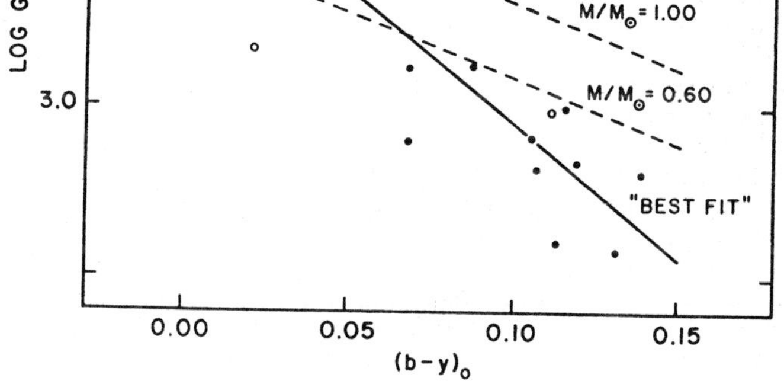

Fig. 4. Log g is plotted against $(b-y)_{0}$ for HB A-stars; the values are derived from Fig. 2. Also shown is an "eyeball" best-fit line and two lines of constant mass.

"best fit" line, the points scatter around the line with a standard deviation of ±0.18 for the field stars, ±0.34 for the cluster stars, and ±0.22 for all stars taken together. The "best fit" line should be parallel to the constant mass lines, since, as explained in the introduction, the HB A-stars can all be expected to have nearly the same mass. Further, as explained in the introduction and in Section 5, the mean mass of these stars is expected to be near 0.6 $M_\odot$. Again, the fact that the "best fit" line has a steeper slope than the constant-mass lines is what is defined as the "discrepancy".

4. POSSIBLE CIRCUMSTELLAR SHELLS

In this section we consider the possibility that the "discrepancy" is due to circumstellar shells, such that the stars are more highly reddened than one would deduce from the standard methods. These stars have quickly evolved from the helium flash stage, during which a mass loss of about 0.2 $M_\odot$ occurs (Renzini 1981), and one could conceive of the shell still being dense enough to "redden" the star. The first question is whether the "discrepancy" can be "explained" by reddening. In the case of the hot stars, the answer is no. The reddening line does copy not carry them in the right direction, as can be seen in Figure 2, and it would take a "negative" reddening of 0.04 mag. to explain the hot end of the discrepancy. In the case of the cooler stars, the answer is yes. The reddening needed is E_{B-V} about 0.022 mag.

The second question is, if the circumstellar shell can explain the discrepancy in the cool stars, does it make sense physically? The answer is no. Globular cluster red giants are observed to be losing mass at a rate of about 2×10^{-9} $M_\odot$/yr with an outflow velocity of about 10 km/sec (Cohen 1976). Furthermore, the mass loss occurs only near the tip of the giant branch (Renzini 1981), and thus the mass loss is cut off quickly after the helium flash. We assume that the star loses its 0.2 $M_\odot$ at a constant rate of 2×10^{-9} $M_\odot$/yr. This means that the mass loss occurs over an interval of $\sim 10^8$ years. At a constant rate of mass loss the density in the shell is given by $\dot{M}/4\pi t r^2$, where $\dot{M}$ is the mass loss rate, t is the distance the gas flows in unit time at the outflow velocity, and r is the radial distance from the center of the star. If we consider a position at 100 $R_\odot$, we calculate a density of about 10^8 H atoms/cm^3, which is in good agreement with observed shell densities (Reimers 1975). The integral of the density from an inner shell radius r_o to infinity is $M = \dot{M}/4\pi t r_o$. At 10 km/sec the gas can travel 10 pc in 10^8 years, so the shell relatively quickly becomes "infinite". The total mass in a 1 cm^2 column from r_o = 100 $R_\odot$ to infinity is about 10^{-3} gm. One magnitude of interstellar extinction is produced by about 10^{-3} gm of the interstellar medium (Woolf 1973), so we find that a shell can produce up to 1 magnitude of extinction during the active mass loss phase. What happens after 10^6 years, when the star has settled on the ZAHB? In 10^6 years the gas travels a distance of 3×10^{19} cm. The mass in a 1 cm^2 column from an inner radius of 3×10^{19} cm to infinity is about 4×10^{-10} gm, which is far too small to have any significant extinction. We conclude that, by the time the star settles on the ZAHB, the shell has dissipated. Thus, circumstellar shells can not explain the "discrepancy", even for the cooler stars.

5. COMPARISON WITH EVOLUTIONARY MODELS

An early objective of the comparison of these observations with evolutionary models was to make a critical test of current evolutionary theory. In view of the "discrepancy" discussed earlier, our present objective is to use this comparison to evaluate the model atmospheres used to derive the effective temperatures and surface gravities. Two figures are discussed in this section. Surface gravity vs. log T_{eff} is shown in Figure 5. Values are shown which are derived by fitting our scans to ATLAS models as follows: small solid dots - field stars observed at Palomar; + signs - M 5 stars; small x's - M 13 stars; large open circles - M 92 stars. Two field stars were observed both at Palomar and on NOAO telescopes; the Palomar and NOAO values are connected by a line in each case. Also shown are ZAHB models by Sweigart and Gross (1976). At the left end of each line is the helium abundance; at the right end of each line is the core mass. The small numbers give the total stellar mass in solar units; the position of the number along the line indicates the position of a star of that mass. Note that in the log g - T_{eff} plane isochrones lie very close to the ZAHB's. We see that the semi-empirical points show a slope which is steeper than the evolutionary sequences.

In a previous discussion (Hayes and Philip 1979) a preferred value of the helium abundance, Y = 0.25, and of the core mass, $M_c = 0.475\ M_\odot$, has been derived. With these values the models cross the distribution of semi-empirical points at a position corresponding approximately to $0.61\ M_\odot$.

A theoretical HR diagram is shown in Figure 6. The semi-empirical points are coded as in Figure 5. Also shown are ZAHB's (dashed lines) and evolutionary tracks (solid lines) from Sweigart and Gross (1976). The numbers on the left indicate the helium abundance; the core mass for all members is $0.475\ M_\odot$. The smaller numbers give the position and value of the total stellar mass. The observations fall in a region which corresponds to a total stellar mass of approximately $0.61\ M_\odot$, with Y = 0.25.

Another way to determine masses for horizontal-branch stars is to apply pulsation theory to double-mode RR Lyrae variables (Cox, King and Hodson 1980, Cox, Hodson and Clancy 1983, Cox 1984). The results of this process give masses of $0.65 \pm 0.05\ M_\odot$ for ten stars in M 15 (an Oosterhoff II cluster) and $0.55\ M_\odot$ for two stars in M 3 (an Oosterhoff I cluster) (Cox, Hodson and Clancy 1984). These values are in good agreement with those found here.

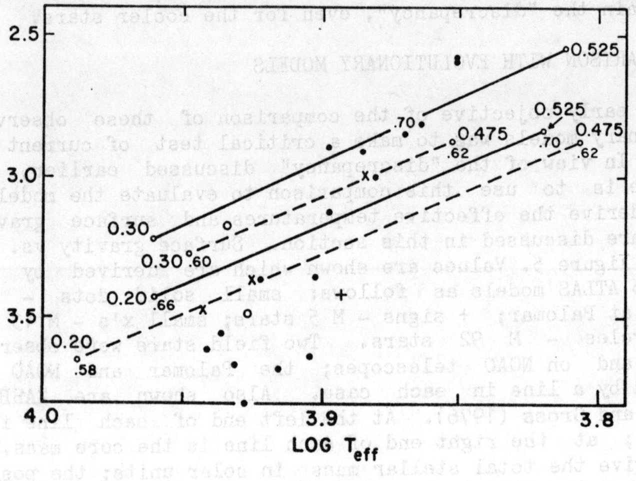

Fig. 5. Log g is plotted against log T_{eff} for Field HB A-stars (filled small circles) and globular cluster HB A-stars in M 5 (crosses), M 13 (x's) and M 92 (larger open circles). Also shown are ZAHB models by Sweigart and Gross (1976).

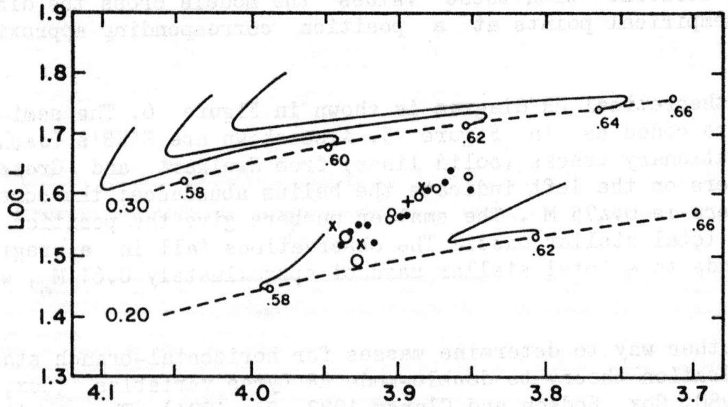

Fig. 6. A theoretical HR diagram. Theoretical ZAHB's (dashed lines) and evolutionary tracks (solid lines) from Sweigart and Gross (1976) are shown, along with points representing HB A-stars, coded as in Fig. 5.

CONCLUSIONS

To summarize the results: scans of horizontal-branch A-stars have been fitted to ATLAS model atmospheres in order to derive effective temperatures and surface gravities. The values which are obtained are used to calculate semi-empirical masses. These masses are compared to evolutionary tracks; in the mean there is good agreement. The mean mass is also in good agreement with the results of applying pulsation theory to observations of double-mode RR Lyrae stars in globular clusters.

The surface gravities so obtained do show a "discrepancy", however, in that the cooler stars show low values of the surface gravity which are equivalent to unbelievably low values of the derived mass, and that the hotter stars show high values of the surface gravity which are equivalent to unbelievably large values of the derived mass. The possibility that interstellar or circumstellar reddening could cause the "discrepancy" is examined and rejected.

The possibility that the ATLAS models have a fundamental problem is shown not to be the case by fitting scans of Population I stars to the models; the temperatures and surface gravities so obtained are in good agreement with those found by more fundamental methods. The small systematic shift found in the surface gravities for Population I stars may be explained easily by minor problems with the models, a systematic error in the absolute calibration of the energy distribution of Vega, the spectrophotometry of the sample of stars, inhomogeneities in the sample of stars with respect to evolutionary states and metallicity of the stars, and possible mismatches in the central wavelength and bandpass of the spectrophotometry with respect to the model atmosphere energy distributions.

Since the "discrepancy" does not appear to be present for Population I stars, it must be a physical effect relating to the lower surface gravity or lower metallicity of the HB A-stars. Since the ATLAS models assume LTE, and the line lists were selected for Population I stars, we consider NLTE and (perhaps) a lack of suitable line lists to be the cause.

With appropriate model atmospheres, semi-empirical masses can be derived for horizontal-branch A-stars. The precision of the masses is low for individual stars, but would be high enough for a group of stars to make a critical test of evolutionary theory possible. The precision is lower than for the case of double-mode RR Lyrae stars, but the scarcity of the latter argues of the value of using horizontal-branch A-stars as a supplemental way to obtain masses.

REFERENCES

Breger, M. 1976a Astrophys. J. Suppl. 32, 1.
Breger, M. 1976b Astrophys. J. Suppl. 32, 7.

Cohen, J. G. 1976 Astrophys. J. Letters 203, L127.
Cox, A. N. 1984 In IAU Symposium No. 105, Observational Tests of the Stellar Evolution Theory, A. Maeder and A. Renzini, eds., Reidel, Dordrecht, p. 421.
Cox, A. N., King, D. S. and Hodson, S. W. 1980 Astrophys. J. 236, 219.
Cox, A. N., Hodson, S. W. and Clancy, S. P. 1983 Astrophys. J. 266. 94.
Hayes, D. S. 1978 in IAU Symposium No. 80, The HR Diagram, A. G. D. Philip and D. S. Hayes, eds., Reidel, Dordrecht, p. 65.
Hayes, D. S. 1985 in IAU Symposium No. 111, Calibration of Fundamental Stellar Quantities, D. S. Hayes, L. E. Pasinetti and A. G. D. Philip, eds., Reidel, Dordrecht, p. 225.
Hayes, D. S. and Latham, D. W. 1975 Astrophys. J. 197, 593.
Hayes, D. S. and Philip, A. G. D. 1979 Publ. Astron. Soc. Pacific 91, 71.
Hayes, D. S. and Philip, A. G. D. 1981 in IAU Colloquium No. 68, Astrophysical Parameters for Globular Clusters, A. G. D. Philip and D. S. Hayes, eds., L. Davis Press, Schenectady, p. 409.
Hayes, D. S. and Philip, A. G. D. 1983 Astrophys. J. Suppl. 53, 759.
Hayes, D. S. and Philip, A. G. D. 1986 in preparation.
Kurucz, R. L. 1981 private communication.
Oke, J. B. and Gunn, J. 1983 Astrophys. J. 266, 713.
Lester, J. B., Love, M. C. and Kurucz, R. L. 1982 Astrophys. J. 260, 272.
Philip, A. G. D. 1978 in IAU Symposium No. 80, The HR Diagram, A. G. D. Philip and D. S. Hayes, eds., Reidel, Dordrecht, p. 209.
Philip, A. G. D. 1983 private communication.
Philip, A. G. D. 1985 in Horizontal-Branch and UV-Bright Stars, A. G. D. Philip, ed., L. Davis Press, Schenectady, p. 41.
Philip, A. G. D., Cullen, M. and White, R. E. 1976 Dudley Observatory Report No. 11.
Reimers, D. 1975 in Problems in Stellar Atmospheres and Envelopes, B. Baschek, W. Kegel and G. Traving, eds., Springer Verlag, Heidelberg, p. 319.
Renzini, A. 1981 in IAU Colloquium No. 59, Effects of Mass Loss on Stellar Evolution, C. Chiosi and R. Stalio, eds., Reidel, Dordrecht, p. 319.
Schild, R. E. 1977 Astron. J. 82. 337.
Sweigart, A. V. and Gross, P. G. 1976 Astrophys. J. Suppl. 32, 367.
Woolf, N. J. 1973 in IAU Symposium No. 52, Interstellar Dust and Related Topics, J. M. Greenberg and H. C. van de Hulst, eds., Reidel, Dordrecht, p. 485.

THE FIELD HORIZONTAL-BRANCH STAR HD 109995: NEW RESULTS WITH COADDED ULTRAVIOLET AND OPTICAL REGION SPECTRA

Saul J. Adelman[1,2,3,4] and David S. Leckrone[2]

Laboratory for Astronomy and Solar Physics
NASA Goddard Space Flight Center

ABSTRACT. A comprehensive ultraviolet and optical region abundance analysis of the field horizontal-branch Population II A-type star HD 109995 is described. Coaddition of IUE high-dispersion images and DAO 6.5 Å/mm IIaO spectrograms improved the signal-to-noise ratio of the data. We have identified ultraviolet lines whose analysis will provide more complete and accurate elemental abundances than those obtained from optical region spectra alone. A preliminary elemental abundance analysis of the optical region shows that log $Z/Z_\odot$ = -2. A first attempt to synthesize two Fe II ultraviolet resonance lines yields an iron abundance a few tenths of a dex higher that the average obtained from optical region Fe II lines.

1. INTRODUCTION

HD 109995 is one of the prototype field horizontal-branch Population II A-type stars. From a study of low dispersion IUE exposures, Böhm-Vitense (1980) found its mass is about 0.5 $M_\odot$ and its temperature 8100 K, with these values somewhat dependent on the exact metallicity. Abundance analyses (e.g., Wallerstein and Hunziker 1964, Kodaira et al. 1969, Kodaira 1973) show that it is metal poor (log $Z/Z_\odot$ = - 1.5) with the anomalies generally increasing with atomic number. In the optical only lines of Na I, Mg I, Mg II, Al I, Si I, Si II, Ca I, Ca II, Sc II, Ti II, V II, Cr I, Cr II, Mn I, Fe I, Fe II, Sr II, Y II and Ba II have been studied. But the derived abundances, except for those from Ti II, Fe I and Fe II lines, depend on at most two lines per atomic species.

---

[1] NRC-NASA Research Associate.
[2] Guest Investigator with the International Ultraviolet Explorer Satellite which is sponsored and operated by NASA, by the Space Research Council of the United Kingdom, and by ESA; Guest Investigator at the Regional Data Analysis Facility at NASA Goddard Space Flight Center.
[3] Visiting Observer at Dominion Astrophysical Observatory
[4] Visiting Astronomer, Kitt Peak National Observatory

A. G. Davis Philip, (ed.)
Horizontal-Branch and UV-Bright Stars 75-83

To improve our knowledge of the abundances of HD 109995, we have undertaken the first high dispersion ultraviolet analysis with IUE satellite data of a Population II field horizontal-branch A star and a new photographic region analysis. Further, Adelman with D. S. Hayes and A. G. Davis Philip have observed C I, N I and O I lines in the infrared with CCD detectors in the coude' spectrograph of Kitt Peak National Observatory. The O I triplet λλ7772-7775 is definitely present. From these observations C, N and O abundances (or upper limits) will be derived consistent with our forthcoming ultraviolet and optical study. An earlier study by Kodaira and Tanaka (1972) found that C and O are at least deficient by a factor of ten.

2. ULTRAVIOLET LINE IDENTIFICATIONS

We have obtained eight SWP and six LWP high dispersion IUE images of HD 109995. Two exposure times were used with each camera to increase the region of good exposure. The images for each camera were added as described by Adelman and Leckrone (1985) to increase the signal-to-noise ratio.

Longward of λ1500, the ultraviolet energy distribution of HD 109995 is fairly flat in F_λ. But shortward of this wavelength it rapidly decreases in brightness. As the sensitivity of the IUE camera-detector combinations increase towards the longward portions of their spectral ranges, the primary regions for our line identification study are λλ1400-2100 and λλ2450-3050. Some, especially strong, lines can be distinguished from noise shortward of these regions.

In the ultraviolet, the line density is much greater than the optical region. But it is not as great as, nor are the lines as strong as, those in θ Leo (A2V), the coolest Population I star for which we have coadded many IUE high dispersion spectra. Figures 1 through 3 show selected regions of the uv spectrum. Figure 1 shows λλ2590-2602 where there are strong Fe II lines, Figure 2 λλ1740-1750 with two strong N I lines and Figure 3 λλ1655-1660 with C I multiplet 2.1

Our primary atomic line identification sources were The Ultraviolet Multiplet Table (Moore 1950, 1952, 1962), the finding list for Moore's newer multiplet tables (Adelman et al. 1985), A Multiplet Table of Astrophysical Interest (Moore 1945) for longward of λ3000, NSRDS-NBS 68, part I (Reader and Corliss 1980) and studies of particular atomic species: Ti II (Huldt et al. 1982), Cr II (Johnsson, private communication) and Fe II (Johansson 1978). There are still many lines without good identifications. The majority of identified lines are those of Fe II. The following atomic species have lines which we are certain are present in our IUE spectra: C I, C II, N I, O I, Mg I, Mg II, Al II, Al III, Si I, Si II, P II, S I, Cl I, Ti II, V II, Cr II, Mn II, Fe I, Fe II, Co II, Ni II, Zn II and Ge II. We expect both to improve the accuracy of abundances based on only a few optical lines and to derive abundances of additional elements. For a few atomic species comparison of resonance lines and those arising from higher lying lower

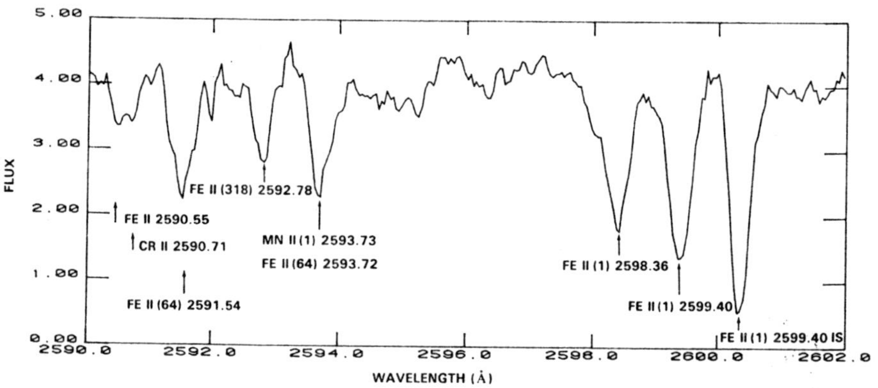

Fig. 1. The λλ2590-2602 region of HD 109995. The strongest identified lines are those of Fe II. The interstellar (IS) component of Fe II (1) 2599.40 is shifted longward by 1.14 Å as the stellar radial velocity is -132 km s^{-1}. This spectrum is the coaddition of three LWP images.

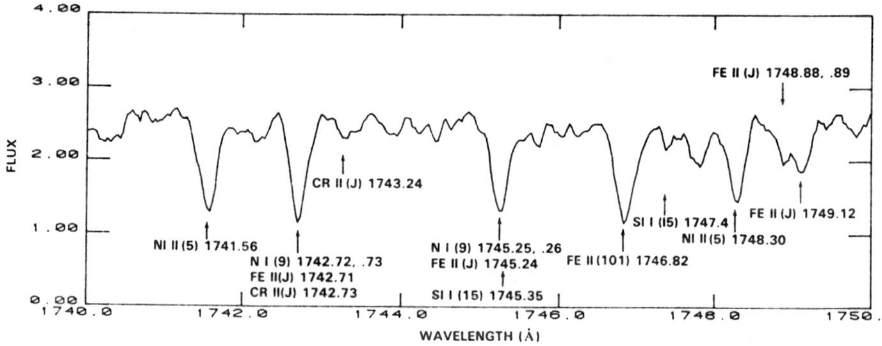

Fig. 2. The λλ1740-1750 region of HD 109995. The two lines of N I multiplet 9 are blended with weaker lines of other elements. To deduce the nitrogen abundance, the abundances of these other elements must also be determined. Also seen are two lines of Ni II multiplet 5. The Fe II lines with a J for their multiplet numbers are from Johansson (1978). This spectra is the coaddition of five SWP images.

levels, might yield information on possible non-LTE effects.

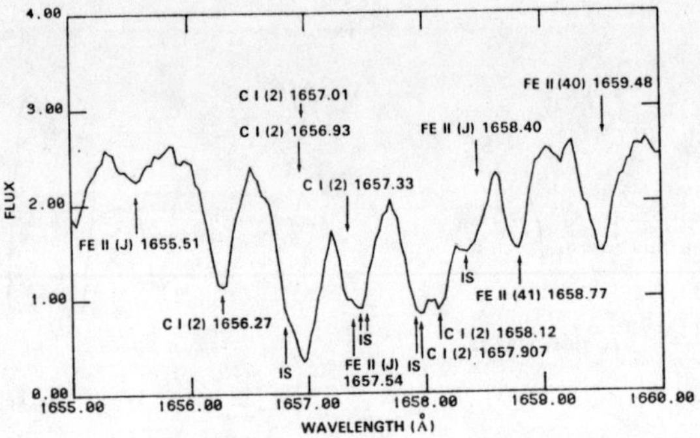

Fig. 3. The λλ1655-1660 region of HD 109995. The lines of C I multiplet 2 and Fe II lines are identified. The positions of the interstellar (IS) components of the C I lines are indicated.

3. OPTICAL REGION SPECTRA

There are considerable uncertainties in the published equivalent widths of the metal lines of HD 109995 in the literature (e.g., Wallerstein and Hunziker 1964 vs. Kodaira et al. 1969). To resolve these differences and to produce an improved set of equivalent widths, 15 6.5 Å/mm IIaO spectrograms of HD 109995 were obtained at the Dominion Astrophysical Observatory in March 1985. Using the procedures discussed in Hill and Adelman (1985), they were added together to increase the signal-to-noise ratio. The S/N of the coadded spectrum is estimated conservatively to be 30. The equivalent widths were measured by fitting Gaussian profiles through the metal lines. The radial velocity of -132 km/s is the same as that of Kodaira et al. (1969). However, we find v sin i = 17 ± 1 km/s compared with 27 km/s found by Peterson et al. (1983). The IUE data are consistent with the former, but not the latter value.

The optical lines were identified primarily by use of A Multiplet Table of Astrophysical Interest (Moore 1945). Certain weak lines included in previous studies were not positively identified, in particular the V II and Ba II lines identified by Wallerstein and Hunziker (1964) and the Mn I, Y II and Ba II lines identified by Kodaira et al. (1969). The equivalent widths of the former authors (WH) tend to be smaller than those of this study while those of the latter authors (KGO) tend to be larger. This study's values are close to that average of these two previous studies. The least squares relations for W_λ in mÅ are as follows:

$$W_\lambda(\mathrm{DAO}) = 1.25\ W_\lambda(\mathrm{WH}) + 2.73$$
and
$$W_\lambda(\mathrm{DAO}) = 0.81\ W_\lambda(\mathrm{KGO}) \ -6.08$$

The H_γ profile published by Kodaira et al. (1969) is in good agreement with that derived from the coadded DAO spectrograms. The line shoulders are well-matched with the coadded H_γ profile being slightly shallower in the core and deeper in the wings. The latter is an effect of a better signal-to-noise ratio.

4. CHOICE OF MODEL AND PRELIMINARY RESULTS

The criteria to be used in the selection of a final model atmosphere for HD 109995 are the match of the model's predictions with the ultraviolet and optical region fluxes, especially the Balmer jump region and H_γ profile; equal abundances from Fe I and Fe II lines; and the model having the same metallicity as the star. Böhm-Vitense (1980) showed that the ultraviolet fluxes would be reasonably fit by an 8000 K, log g = 3.0 model with log $Z/Z_\odot$ = -1.5.

For an initial estimate of the iron abundance and microturbulent velocity, we selected 8000 K and 8500 K, log g = 3.0 and 3.5 models with log $Z/Z_\odot$ = -1 and -2 from Kurucz (1979). The Fe I and Fe II gf-values are from the critical compilation by Martin, Fuhr and Wiese (1986). The Fe II damping constants were calculated from the sums given by Kurucz (1981) and those for Fe I are semi-classical approximations. Microturbulent velocities are found by requiring that there be no dependence of the deduced Fe abundance on equivalent width. Both optical and ultraviolet lines indicated that [Fe/H] = -2 in accord with the results of Kodaira (1973).

Comparison of the predictions of log $Z/Z_\odot$ = -1.5 and -2.0 model atmospheres calculated with the ATLAS6 code (Kurucz 1979) with optical spectrophotometry over λλ3400-6790 of Philip and Hayes (1983) indicates 8250 K, log g = 3.40. However, the predicted fluxes do not agree well with those derived from IUE low dispersion exposures. The uv fluxes and H_γ profile suggest a cooler effective temperature of order 300 K. The log $Z/Z_\odot$ = -1.5 model fluxes agree slightly better with the uv fluxes than those of the log $Z/Z_\odot$ = -2.0 model. This indicates that the model atmospheres are underblanketed compared to the star.

As the log $Z/Z_\odot$ = -1.5 and -2.0 model atmospheres which fit the optical fluxes are very similar, we decided to use our log $Z/Z_\odot$ = -2.0 model for the abundance analysis. A microturbulence of 1.4 ± 0.1 km/s was found from analysis of the Fe I and Fe II lines. In Table I we give the values of log N/H and [N/H] = log N/H -(log N/H)$_\odot$ for HD 109995, the values of log N/H for the Sun and the average log N/H for ten Population I B and A stars analyzed with the same gf values (Adelman 1985). The differences between the Ca I and Ca II and the Cr I and Cr II results are also seen in the Population I stars. The latter difference may well be due to the poor quality of the Cr II gf-values. The abundance anomalies for the most part agree quite well with those of Kodaira

(1973) who found HD 109995 more metal poor than most other researchers. The major differences are that both calcium and chromium are more deficient in the present analysis despite the differences between the neutral and singly-ionized results.

TABLE I

Comparison of Abundances

Atomic Species	HD 109995 Log N/H	[N/H]	Sun Log N/H	Pop I Stars Log N/H
Mg I	-5.90	-1.52	-4.38	-4.23
Mg II	-5.44	-0.96	-4.38	-4.29
Al I	-8.00	-2.49	-5.51	-5.81
Si II	-5.84	-1.47	-4.37	-4.40
Ca I	-7.48	-1.82	-5.66	-5.71
Ca II	-7.74	-2.18	-5.66	-5.96
Sc II	-10.87	-1.91	-8.96	-9.16
Ti II	-8.32	-1.30	-7.02	-6.97
Cr I	-8.44	-2.56	-5.88	-6.05
Cr II	-7.57	-1.69	-5.88	-5.78
Fe I	-6.51	-2.14	-4.37	-4.30
Fe II	-6.24	-1.87	-4.37	-4.37
Sr II	-11.66	-2.46	-9.10	-8.88

5. ULTRAVIOLET LINE SYNTHESIS

We have undertaken fully self-consistent abundance analyses of nine stars, including HD 109995, six normal late B and early A-type dwarfs and two Bp HgMn stars, which are based on the combination of high signal-to-noise optical region and IUE spectra. The rationale and methodology are briefly described in Leckrone and Adelman (1985). The techniques and problems associated with the detailed synthesis of the complex line blends found in IUE spectra are discussed in Leckrone (1984). Our initial efforts have focused on the synthesis of two resonance features of Fe II, λ2607.09 and λ2625.7. The log gf-values of these lines are particularly well known (to about 10%). Moreover, since the optical region spectrum of iron is rich and well understood, we can immediately establish whether the IUE data yield quantitative abundances consistent with ground-based observations. Preliminary results for the normal stars indicate agreement between λ2607 and the average optical iron abundance to within about ± 0.1 dex, while λ2626 typically yields values higher than the optical average by about 0.25 dex. Since abundances derived from optical region spectra alone are usually uncertain by at best 0.2 dex, the agreement between IUE and optical region data can be viewed as reasonably good, giving confidence that the rich ultraviolet spectra will yield reliable and unique abundance information.

Figure 4 illustrates the preliminary synthesis of Fe II λ2607 in HD 109995. The calculations utilized the model atmosphere,

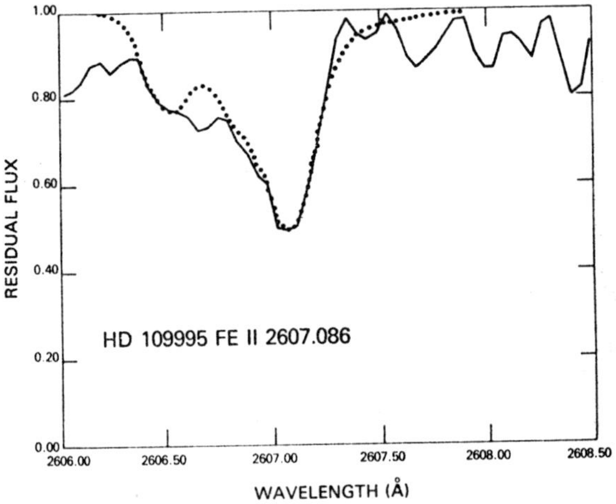

Fig. 4. The profile of the ultraviolet resonance line of Fe II at λ2607.09 in HD 109995. The solid curve portrays the observations, while the dotted curve represents the computed profile with log Fe/H = -6.0. The computations include approximately 55 relatively weak contaminant blending lines, but is dominated by λ2607.

microturbulent rotation and velocity described in Sections 3 and 4. Abundances of contaminant blending species (principally Ti, V, Cr, Mn and Co) were either adopted from Kodaira (1973) or were taken as 2 dex below solar. The observational data are based on a coaddition of only 3 of our 6 LWP high dispersion images. When we incorporate our recently obtained data, the signal-to-noise ratio of the observations will be somewhat improved. The best fit to λ2607 was obtained with log Fe/H = -6.0, while log Fe/H = -5.6 was obtained for λ2626. These are somewhat higher than the average value, -6.2, obtained from optical Fe II lines. No great significance should yet be attached to these differences, however, pending further refinement of the model atmosphere, development of the final coadded IUE spectra, and analysis of many more ultraviolet lines. It is important to note that the problems of signal-to-noise and the resolution of blends which make analyses of the IUE spectra very difficult will be greatly ameliorated by the Hubble Space Telescope High Resolution Spectrograph. It is only with the latter instrument that moderately weak ultraviolet lines will be observed with sufficient accuracy to allow their reliable analysis.

ACKNOWLEDGMENTS

SJA thanks Dr. Graham Hill and Mr. Wesley Fisher for their help with the coaddition of the DAO spectra of HD 109995. This research was supported in part by NASA grant NAG 5-218 to the Citadel.

REFERENCES

Adelman, C. J., Adelman, S. J., Fischel, D. and Warren, W. H. 1985 Astron. Astrophys. Suppl. 60, 339.
Adelman, S. J. 1985 Astron. Astrophys. Suppl., in press.
Adelman, S. J. and Leckrone, D. S. 1985 in preparation.
Böhm-Vitense, E. 1980 Astrophys. J. 243, 213.
Hill, G. and Adelman, S. J. 1985 in IAU Colloquium No. 90, Upper Main Sequence Stars with Anomalous Abundances, C. R. Cowley and C. Megessier, eds., Reidel, Dordrecht, in press.
Huldt, S., Johansson, S., Litzen, U. and Wyart, J. -F. 1982 Physica Scripta 25, 401.
Johansson, S. 1978 Physica Scripta, 18, 217.
Kodaira, K. 1973 Astron. Astrophys. 22, 273.
Kodaira, K., Greenstein, J. L. and Oke, J. B. 1969 Astrophys. J. 155, 525.
Kodaira, K. and Tanaka, K. 1972 Publ. Astron. Soc. Japan 24 355.
Kurucz, R. L. 1979 Astrophys. J. Suppl. 40. 1.
Kurucz, R. L. 1981 SAO Special Report 390.
Leckrone, D. S. 1984 Astrophys. J. 286, 725.
Leckrone, D. S. and Adelman, S. J. 1985 Bull Am. Astron. Soc. 16, 973.
Martin, G. A., Fuhr, J. R. and Wiese, W. L. 1986 in preparation.
Moore, C. E. 1945 A Multiplet Table of Astrophysical Interest, Princeton University Observatory.
Moore, C. E. 1950 An Ultraviolet Multiplet Table, NBS Circular No. 488, Section I.
Moore, C. E. 1952 An Ultraviolet Multiplet Table, NBS Circular No. 488, Section II.
Moore, C. E. 1962 An Ultraviolet Multiplet Table, NBS Circular No. 488, Sections III-V.
Peterson, R. C., Tarbell, T. D. and Carney, B. W. 1983 Astrophys. J. 265, 972.
Philip, A. G. D. and Hayes, D. S. 1983 Astrophys. J. Suppl. 53, 751.
Reader, J. and Corliss, C. H. 1980 Wavelengths and Transition Probabilities for Atoms and Atomic Ions, NSRDS-NBS 68, part I.
Wallerstein, G. and Hunziker, W. 1964 Astrophys. J. 140, 214.

DISCUSSION

LECKRONE: Something that I had not appreciated is how metal-rich some of the horizontal-branch stars are. You gave, Ken, as your canonical scale -0.7.

JANES: That is the most metal-rich at the moment.

PHILIP: Those clusters don't have any blue horizontal-branch stars.

LECKRONE: Now, what is the range, how weak do they become?

JANES: You are down by a factor of 50, or a factor of 100 from the Sun.

LECKRONE: We started out with HD 109995 using a grid of log $Z/Z_\odot = -2$ models because that was what Kodaira, Greenstein and Oke got and this is appproximately the correct metallicity.

JANES: Anything as metal-rich as -1.0 will not have blue horizontal-branch stars in it. If you are forced to bring this star above -1.0 there would be a question of its relation to globular cluster stars.

ADELMAN: HD 161817 is more metal-rich by about 0.3 dex relative to HD 109995. The analysis of HD 109995 uses latest oscillator strength computations and a lot of weak lines, lines down to 10 milliangstroms.

JANES: Many people have got caught in the same trap. They decided that in order to avoid problems of dealing with detailed atmospheres they would take the weak lines. Then they got caught by the problem of where to set the continuum. All you have to do is to estimate the level of the continuum off by the tiniest hair and the abundances will change a great deal.

PHILIP: Drawing the continuum through areas containing noise and no lines is the correct way of doing it and I think that some of the errors made earlier were made by people who did not use the noise level but were sort of topping the peaks.

JANES: In red giants you are sort of guessing how many missing weak lines there are. I gave a student a spectrum to do and he came back to me some time later and he said "I can't do it; I can't tell where to put the continuum.".

ADELMAN: Ken, I will check my fine analysis of FHB stars by synthesising parts of their spectra. This should remove any questions. I believe the continuum level is reasonably well determined, but there is no reason not to check this point.

THE ROTATION OF HORIZONTAL-BRANCH STARS

Ruth C. Peterson

Whipple Observatory, Smithsonian Institution

ABSTRACT. The current empirical status of the presence of primordial stellar rotation among blue horizontal-branch stars in globular clusters is outlined. Spectroscopic approaches are sketched towards establishing the influence of this rotation on the color of a cluster's horizontal branch, the determination of its age, and the occurrence of mixing and mass loss among its giants.

From the broadening of spectral lines observed with the echelle spectrograph on the Multiple Mirror Telescope at a resolution of 10 km/s, the presence of rotation among blue horizontal-branch (BHB) stars has been inferred now in four out of five globular clusters studied, namely M 3, M 5 and M 13 (Peterson 1983) and NGC 288 (Peterson 1985b). Only M 4 shows no rotation at the nominal detection limit of v sin i > 15 km/s (Peterson 1985a). The maximum v sin i value seen to date is about 30 km/s; eleven out of 33 stars surveyed show a v sin i > 15 km/s. Taken together, these facts imply that the rotation was present within the star during its lifetime on the main sequence, rather than induced subsequently through spinup by a companion (Peterson 1983).

As seen in Table I, taken from Peterson (1985b), these results also suggest strongly that rotation influences the mean HB color of globular clusters of intermediate metallicity. For the ranking of the clusters in order of increasing rotation is also a ranking in order of increasing B/B+R, the ratio of BHB stars to all HB nonvariables.

Because factors other than rotation may also be involved, the sample of five clusters is too limited to extrapolate the influence of rotation to clusters with B/B+R = 0, and thus solve the second-parameter problem. Rood (1973) and Rood and Crocker (1985a and b) have shown that the abundance of the CNO elements, the helium abundance, and age all could affect B/B+R as well. Unfortunately, it is virtually impossible to measure or infer the abundance of helium in any system, and it is very difficult to set limits on more than one of the remaining parameters in the metal-poor systems where red HB color predominates, because they are so distant. For example, only the age has been determined in the extreme second-parameter case of the dwarf spheroidal galaxy Draco, for which B/B+R = 0.13 (Baade and Swope 1961) yet [Fe/H] ~

A. G. Davis Philip, (ed.)
Horizontal-Branch and UV-Bright Stars 85–89

TABLE I

HORIZONTAL-BRANCH COLOR VERSUS ROTATION AND OXYGEN ABUNDANCE IN FIVE GLOBULAR CLUSTERS

Parameter	M 4	M 3	M 5	NGC 288	M 13
HB Color Ratio B/B+R	0.44	0.55	0.76	1.00	1.00
Rotation:					
Number of stars	9	4	7	7	6
Maximum v sin i (km/s):					
Highest	14 ± 4	18 ± 3	20 ± 3	22 ± 3	33 ± 4
Second	14 ± 4	11 ± 3	16 ± 4	19 ± 3	29 ± 4
Average v sin i (km/s)	9.6 ± 4	12.0 ± 2.1	10.4 ± 2.4	14.4 ± 2.6	21.0 ± 4.1
Fraction of stars with v sin i >:					
10 km/s	0.67	0.75	0.67	0.86	1.00
15 km/s	0.00	0.25	0.29	0.57	0.67
20 km/s	0.00	0.00	0.14	0.14	0.67
Oxygen Abundance:					
[O/Fe]	...	+0.3	+0.1	+0.3	-0.3
[Fe/H]	...	-1.6	-1.1	-1.2	-1.4
[O/H]	-0.9	-1.3	-1.0	-0.9	-1.7

- 2 (Stetson 1984). A young age has been found unlikely, from photometry to the main-sequence turnoff by Carney and Seitzer (1985) and Stetson, Vandenberg and McClure (1985). Draco is too remote to measure directly either BHB rotation or the CNO abundance via the O I line at 6300.3 Å (see Kraft 1985 for a discussion of the latter).

In view of the fact that rotation has been detected in four of five globular clusters studied, and influences the position of a star on the HB, it seems crucial to try to establish the influence of rotation on several other phases of metal-poor stellar evolution. Of vital importance is the effect on the age determination of globular clusters from the luminosity of the main-sequence turnoff (MS TO). This problem, reviewed by Carney (1984), has been recommended as part of a key project for Space Telescope by the Working Group on Stars and Star Clusters in the Galaxy. Theoretically, rotation is expected to prolong the MS lifetime; thus Law (1981) finds that cluster ages might be underestimated by 30% if rotation is actually present but ignored.

An observational test is possible through accurate CCD TO photometry in a cluster in which rotation is present, since a range of rotational velocities would be expected to spread out stars near the TO. Such CCD photometry is available for two globular clusters. One of these, M 4, is the only cluster whose BHB stars failed to show rotation. Since Richer and Fahlman (1984) have found a very tight, narrow MS TO in M 4, the simplest way to show that rotation has a negligible effect on the MS TO is to detect rotation among the M 4 BHB stars. These are so bright that it is possible to bring down the upper limits on v sin i from 15 km/s to 8 km/s, the level at which rotation has been detected among field BHB stars (Peterson, Tarbell and Carney 1983). Also, hotter M 4 BHB stars should be surveyed at low S/N. These have generally been ignored because they are fainter and their lines are weak, but are theoretically the ones most likely to be rotating rapidly.

Should all this fail to yield detectable line broadening, it would be difficult to escape the conclusion that there is no rotation among M 4 BHB stars, and that some other factor is principally responsible for the fact that Draco has a redder HB but a metallicity at least ten times lower than that of M 4.

Mass loss on the giant branch is probably encouraged by rotation also (Renzini 1977). The only empirical test of this is statistical, namely the correlation (or lack thereof) between the percentage of mass-losing giants in a cluster and its percentage of rotating BHB stars. As with HB color, allowance must be made for other influential factors, in this case luminosity and metallicity.

The suggestion that all mass-losing giants fall within a specific region of the CMD, whatever their metallicity (Peterson 1982), was based primarily on Hα emission as a mass-loss indicator (Cohen 1976). This had to be re-examined in the light of the calculations of Dupree, Hartmann and Avrett (1984) that show that Hα emission can be produced without mass loss, i.e. in a static chromosphere, and that only a blue-shifted Hα core is a signature of mass outflow. Since observations at a resolution of 16 km/s or higher of some bright giants in globular clusters do show such a blue-shifted core, in both Hα and NaD (Peterson 1981), there is still evidence that mass loss does occur in globular clusters. To test its dependence on rotation, additional Hα observations at a resolution of 9 km/s of giants in M 3, M 5 and M 13 were obtained on the 1.5 m telescope at Mt. Hopkins as well as the MMT. The data in hand hint that mass loss does occur somewhat more frequently in clusters with many rotating BHB stars. However, a more thorough survey is needed which includes the luminous giants in all five clusters with BHB observations, plus some extremely metal-poor ones.

According to Sweigart and Mengel (1979), rotation may encourage the mixing of products of the CNO cycle from the interior to the surface of a giant, with consequent carbon depletion and nitrogen enhancement in stars so affected. This suggestion is supported so far by the large amount of both rotation and mixing seen in M 13 compared to the modest amount of both in M 3 (Suntzeff 1981, Peterson 1983). This could be pushed further from spectroscopic observations of CH plus CN or NH in M 4 and NGC 288 giants. Since no rotation was detected in M 4, the carbon depletion exhibited by its giants should not be nearly as dramatic, nor appear at such low luminosities, as in the giants in the high-rotation clusters NGC 288 or M 13.

A correlation between mixing and mass loss might also be established directly. If N I lines are enhanced but C I lines are weakened in rotating field BHB stars compared to nonrotating ones, this would establish the effect. To this end, very high S/N CCD spectra in the vicinity of the infrared N I lines were obtained on the 4m echelle at Kitt Peak for five field BHB stars, but so far the results have been inconclusive. Whether the lines have been detected cannot be said without an accurate dispersion solution, which must wait until a 2-D extraction/reduction package becomes available under IRAF. Accurate stellar temperatures are also required, since these are extremely high-excitation lines. It may be necessary to obtain high S/N data for other high-excitation lines to ensure a reliable nitrogen abundance determination.

To summarize, then, rotation plays an influential role on the HB color of a moderate-metallicity globular cluster, and probably encourages both mixing and mass loss as well. In the not-too-distant future, its effect, if any, on the position of the MS TO, and hence on the derivation of the cluster age, will probably be discerned from highly accurate CCD photometry at the turnoff of a cluster in which rotation has been detected. Likewise, straightforward spectroscopic surveys should soon reveal the extent to which rotation encourages mixing and mass loss. It will be much more difficult to tell whether the lack of rotation is the primary cause of the extremely red HB's of remote globular clusters. A negative answer might be forthcoming if no rotation at all is detected in M 4 after a thorough search of its BHB stars, especially the hottest ones.

REFERENCES

Baade, W. and Swope, H. 1961 Astron. J. 66, 300.
Carney, B. W. 1984 Publ. Astron. Soc. Pacific 96, 841.
Carney, B. W. and Seitzer, P. 1985 preprint.
Cohen, J. G. 1976 Astrophys. J. Letters 203, L127.
Dupree, A. K., Hartmann, L. and Avrett, E. H. 1984 Astrophys. J. Letters 281, L37.
Kraft, R. P. 1985 in Production and Distribution of CNO Elements, ESO Workshop 13-15 May 1985, in press.
Law, W-Y. 1981 Astron. Astrophys. 102, 178.
Peterson, R. C. 1981 Astrophys. J. Letters 248, L31.
Peterson, R. C. 1982 Astrophys. J. 258, 499.
Peterson, R. C. 1983 Astrophys. J. 275, 737.
Peterson, R. C. 1985a Astrophys. J. 289, 320.
Peterson, R. C. 1985b Astrophys. J.Letters 294, L35.
Renzini, A. 1977 in Advanced Stages of Stellar Evolution, P. Bouvier and A. Maeder, eds., Geneva Obs., Geneva, p. 151.
Richer, H. B. and Fahlman, G. G. 1984 Astrophys. J. 277, 227.
Rood, R. T. 1973 Astrophys. J. 184, 815.
Rood, R. T. and Crocker, D. A. 1985a in Production and Distribution of CNO Elements, ESO Workshop 13-15 May 1985, in press.
Rood, R. T. and Crocker, D. A. 1985b in Horizontal-Branch and UV-Bright Stars, A. G. D. Philip, ed., L. Davis Press, Schenectady, p. 99.
Stetson, P. B. 1984 Publ. Astron. Soc. Pacific 96, 128.
Stetson, P. B., Vandenberg, D. A. and McClure, R. D. 1985 preprint.
Suntzeff, N. B. 1981 Astrophys. J. Suppl. 47, 1.
Sweigart, A. V. and Mengel, J. G. 1979 Astrophys. J. 229, 624.

MASS LOSS ON THE BLUE HORIZONTAL BRANCH AND THE ORIGIN OF THE SUBDWARF B STARS

Pierre Demarque and Jo-Ann Eder

Yale University Observatory

Abstract. We have investigated the effects of mass loss on the evolution of blue horizontal-branch stars. We find that mass loss rates in the range $2.5 \times 10^{-10} < \dot{M} < 1 \times 10^{-9}$ $M_{\odot}$/year are needed to explain the existence of the subdwarf B stars.

1. INTRODUCTION

The problem of the evolutionary status of the subdwarf B (sdB) stars continues to puzzle investigators of stellar evolution. Questions abound both about their origin and about their subsequent evolution. For example, do the sdB's form a blue extension of the horizontal-branch (HB), and if so why are they so well represented in the old disk stellar population, which would not be expected to display any kind of blue HB because of its relatively high metallicity? By the same token, why are these stars relatively so rare in globular clusters where low metallicity favors the production of blue HB stars? What is the connection between sdB's and sdO's? Do sdB's evolve into sdO's, or do they represent separate channels of evolution toward the white dwarf stage and, if the latter is correct, does each group evolve into different classes of white dwarfs? The answer to such questions should greatly clarify our understanding of the advanced phases of the evolution of low mass stars. At the same time, old blue stars are believed to be significant contributors to the ultra-violet light of some globular clusters and of elliptical galaxies (Ciardullo and Demarque 1978; Gunn, Stryker and Tinsley 1981). For all these reasons, they are prime candidates for observations with the Hubble Space Telescope.

2. THE INTERNAL STRUCTURE OF THE SUBDWARF B STARS

There are at present two competing models for the internal structure of the sdB stars, neither of which is fully satisfactory. The first model, the "extended HB" model, was first proposed by Greenstein (1971) for the field sdB's and discussed by Faulkner (1972) in terms of stellar interiors. In Faulkner's picture, the sdB

A. G. Davis Philip, (ed)
Horizontal-Branch and UV-Bright Stars 91-97

stars have the same internal structure and evolutionary status as the HB stars, i.e. a helium burning core surrounded by a hydrogen rich envelope. They are thus HB stars which, due to unusually large mass loss, have a hydrogen envelope of only a few percent of a solar mass, and as a result are bluer and slightly fainter than most HB stars. This interpretation is consistent with current theoretical understanding of the HB (Demarque 1973, 1980; Sweigart and Gross 1976), and further seems natural in view of the presence of some sdB's in a least four globular star clusters, in M13 (Sargent and Searle 1971), NGC6752 (Cannon 1981), M15 (Buonnano et al. 1985), and ω Cen (Da Costa et al. 1985). The main drawback of this model is associated with the production of core helium burning stars immediately following the phase of giant evolution, which is difficult to understand in terms of standard evolutionary tracks. Sweigart et al. (1974) point out that unless a substantial amount of mass loss is caused by the helium flash, which seems unlikely in view of Deupree's (1984) recent hydrodynamic studies of the flash, excessive mass loss on the giant branch would force the star to evolve across the HR-diagram to the pre-white dwarf cooling track before helium ignition, thus by-passing the HB, or even the sdB phase.

A possible solution to this problem is that all sdB stars are members of close binary systems in which mass exchange has taken place (Mengel et al 1976). But although there is some observational evidence to support the view that binaries are frequent among sdB's (Graham and Slettebak 1983; Baschek and Norris 1975), the binary hypothesis sets requirements for the formation of sdB's which are probably too infrequently met.

The other model for the sdB stars was suggested more recently By Wesemael et al. (1982). In this model, sdB's have an inert core of carbon which is evolving on a cooling time-scale, and an envelope rich in hydrogen and helium. The coincidence in the HR diagram of the sdB position with the extension of the core helium burning sequence is in this model only fortuitous. The possibility that the sdB's and the redder part of the HB correspond to different evolutionary stages was emphasized by Newell (1973) because of the presence of a gap in the stellar distribution between the sdB's and the HB in the HR diagram. In addition to populating the right region of the HR-diagram, the "cooling carbon core model" has the attractive feature of explaining the sdB to sdO transition as a convective mixing phase along a blueward evolutionary track. Its main drawback, however, is that it cannot be easily understood within the current picture of the evolution of low mass stars. But a post-asymptotic giant branch origin for the sdB stars cannot be ruled out at this point.

3. MASS LOSS FROM BLUE HORIZONTAL-BRANCH STARS

We now wish to describe an attempt to explain the sdB stellar population as an "extended HB". In this paper, mass loss during HB evolution is responsible for the low hydrogen envelope mass. Such a mechanism might provide a natural explanation for the puzzling observation that the sdB population is observed in globular clusters with a particular HB morphology, generally but not always associated with clusters of intermediate metallicity. If indeed mass loss is proportional to metallicity, or CNO abundance, as suggested by the theory of stellar winds (Abbott 1982), only stars in intermediate metallicity (or intermediate CNO abundances) clusters suffer sufficient mass loss in their HB lifetime to be sufficiently affected to evolve away from the asymptotic giant branch and toward the blue. And of course, the most metal rich clusters are known not to have HB stars bluer than the RR Lyrae instability strip. The main drawback of this hypothesis is its adhoc nature since no direct observations of mass loss are available for blue HB stars, and too little is known at this point about the complex interaction of chemical composition, age and other physical parameters such as rotation (Peterson 1983), which determine HB morphology.

The effects of mass loss on HB evolution has been investigated by Laskarides (1973), in an effort to explain observations of period changes with both positive and negative signs among RR Lyrae variables. More recently, Willson and Bowen (1984) have suggested, on the basis of their studies of the dynamics of variable star atmospheres, that mass loss while in the RR Lyrae instability strip modifies the evolution of HB stars in an important way. Since then, Willson and Bowen's (1984) suggestions have been analyzed in some detail by Waugh (1985). In this paper, we shall instead concentrate on the effects of mass loss on HB stars situated on the blue side of the RR Lyrae instability strip.

In view of all these uncertainties, and for the purpose of exploration, we chose to consider mass loss from blue HB stars in the simplest possible way. For two masses at the blue end of the zero-age HB, we constructed a set of evolutionary tracks, each with a constant rate of mass loss. A few trial evolutionary tracks then enabled us to bracket the range of mass loss rates needed to reverse the direction of evolution from redward to blueward.

Calculations were performed for 0.50 and $0.55M_\odot$, and the composition parameters (X,Z) = (0.749, 0.001). The helium core mass was fixed at $M_c = 0.488M_\odot$, consistent with the giant models of Sweigart and Gross (1978). For each mass, a series of evolutionary tracks was constructed assuming the following mass loss rates: 0, 1×10^{-10}, 2.5×10^{-10}, 5×10^{-10}, and 1×10^{-9} $M_\odot$/year. The mass loss is evaluated at each time-step, the mass being removed from the superficial layers of the hydrogen rich envelope. The energy per unit mass in the envelope required to eject the material against the gravitational field of the star is calculated and subtracted from the internal energy of the envelope at each step.

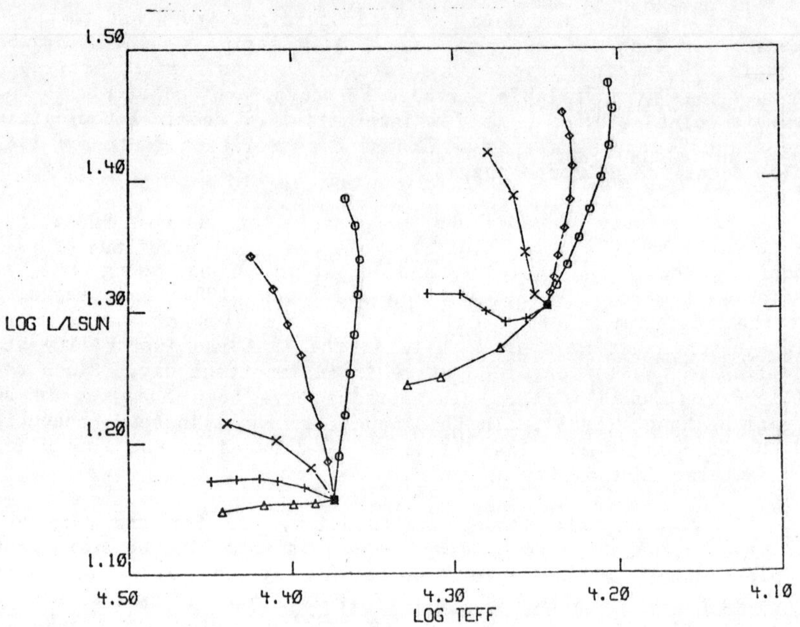

Fig. 1. Evolutionary tracks in the HR diagram for $0.50M_{\odot}$ and $0.55\ M_{\odot}$. For each mass, evolutionary tracks have been constructed for the following mass loss rates (in $M_{\odot}$/year): o, no mass loss; ◊ 1×10^{-10}; × 2.5×10^{-10}; ı 5×10^{-10} and Δ 1×10^{-9}.

TABLE I

Time Since Initiation of Core He Burning (in 10^6 yrs)
$0.5\ M_\odot$

Mass Loss Rate(in $M_\odot$ per yr.)

Model #	0	10^{-10}	2.5×10^{-10}	5×10^{-10}	10^{-9}
2					3.1
4					6.5
6				10.0	10.0
8					13.7
10	17.1	17.1	17.3	17.4	
12				21.5	
14				25.7	
16				29.3	
20	35.8	35.9	36.4		
30	53.2	54.4	54.7		
40	69.4	69.8			
50	84.2	85.1			
60	97.9	99.3			
70	110.2	112.0			
80	119.7				

TABLE II

Time Since Initiation of Core He Burning (in 10^6 yrs)
$0.55\ M_\odot$

Mass Loss Rate(in $M_\odot$ per yr.)

Model #	0	10^{-10}	2.5×10^{-10}	5×10^{-10}	10^{-9}
10	15.8	16.0		15.9	16.1
20	33.3	33.9	33.3	33.5	34.5
25					42.7
30	49.6	49.5		50.3	
40	64.5	64.3	64.8	65.9	
50	78.0	78.0		80.4	
60	90.3	90.5	91.6		
70	101.4	101.9			
80	109.2	108.7	109.7		

The results are illustrated in Figure 1 in the theoretical HR diagram and in Tables 1 and 2. They show clearly that for a mass loss rate in the range $2.5 \times 10^{-10} < \dot{M} < 1 \times 10^{-9}\ M_{\odot}$/year, the direction of evolution is reversed toward the sdB region in the HR diagram.

4. SUMMARY AND DISCUSSION

We have investigated the effects of mass loss on the evolution of blue HB stars. The calculations show that indeed it is possible to reverse the direction of evolution in the HR diagram from redward to blueward, and thus to populate the sdB region of the HR diagram, with mass loss rates in the range $2.5 \times 10^{-10} < \dot{M} < 1 \times 10^{-9}\ M_{\odot}$/year. Although in excess by nearly two orders of magnitudes of the mass loss rates predicted by Abbott (1982) for stars with the relatively modest luminosity of subdwarf stars, the mass loss rates derived here might still be plausible in view of the many uncertainties involved. In particular, the use of Abbott's rates requires a large extrapolation from the rates observed in much more luminous young stars. There is, however, another objection related to the interaction of mass loss with diffusion, which poses a serious problem to our hypothesis. It has been known for a long time that diffusion plays an important role in modifying atmospheric abundances in blue HB stars (Michaud et al 1983). Recent calculations of diffusion rates in the outer layers of sdB's by Michaud and Fontaine (1985) put very stringent limits on the rate of mass loss from these objects, which are several orders of magnitudes lower than required by our hypothesis.

The evolutionary status of the sdB stars thus remain obscure. Detailed spectroscopic observations of these objects in the ultra-violet with the Hubble Space Telescope, with special emphasis on evidence for mass loss and chemical composition anomalies, will be particularly valuable as part of an overall research effort to clarify the nature of old blue stars in the halo and in the disk of our Galaxy.

The numerical work in this paper was performed with the IBM4361 computer system of the Yale Astronomy Department which was purchased with the support of grant AST83-06143 from the National Science Foundation.

REFERENCES

Abbott, D. C. 1982 Astrophys. J. 259, 282.
Baschek, B. and Norris, J. 1975 Astrophys. J. 199, 694.
Buonnanno, R., Corsi, C. E. and Fusi Pecci, F. 1985 Astron. Astrophys. 145, 97.
Cannon, R. D. 1981 in IAU Colloquium No. 68, Astrophysical Parameters for Globular Clusters, A. G. D. Philip and D. S. Hayes, eds., L. Davis Press, Schenectady, p. 501.
Ciardullo, R. B. and Demarque, P. in IAU Symposium No. 80, The HR Diagram, A. G. D. Philip and D. S. Hayes, eds., Reidel, Dordrecht, p. 345.
Da Costa, G. S., Norris, J. and Villumsen, J. V. 1985 preprint.
Demarque, P. 1973 in IAU Colloquium No. 21, Variable Stars in Globular Clusters and in Related Systems, J. D. Fernie, ed., Reidel, Dordrecht, p. 187.
Demarque, P. 1980 in IAU Symposium No. 85, Star Clusters, J. E. Hesser, ed., Reidel, Dordrecht, p. 281.
Deupree, R. G. 1984 Astrophys. J. 287, 268.
Faulkner, J. 1972 Astrophys. J. 173, 201.
Graham, J. A. and Slettebak, A. 1973 Astron. J. 78, 295.
Greenstein, J. L. 1971 in IAU Symposium No. 42, White Dwarfs, W. J. Luyten, ed., Reidel, Dordrecht, p. 46.
Gunn, J. E., Stryker, L. L. and Tinsley, B. M. 1981 Astrophys. J. 249, 48.
Laskerides, P. G. 1973 Monthly Notices Roy. Astron. Soc. 164, 169.
Mengel, J. G., Norris, J. and Gross, P. G. 1976 Astrophys. J. 204, 488.
Michaud, G. and Fontaine, G. 1985 paper presented at workshop on Stellar Remnants, KPNO, Jan. 5 - 11, 1985.
Michaud, G., Vauclair, G and Vauclair S. 1983 Astrophys. J. 267, 256
Newell, E. B. 1973 Astrophys. J. Suppl. 26, 37.
Peterson, R. C. 1983 Astrophys. J. 275, 737.
Sargent, W. L. W. and Searle, L. 1971 Carnegie Yearbook, 70, 404.
Sweigart, A. V. and Gross, P. G. 1976 Astrophys. J. 190, 101.
Sweigart, A. V. and Gross, P. G. Astrophys. J. Suppl. 36, 405.
Sweigart, A. V., Mengel, J. G. and Demarque, P. 1974 Astron. Astrophys. 30, 13.
Waugh, J. F. 1985 Senior Research Project, Yale College.
Wesemael, F., Winget, D. E., Cabott, W., Van Horn, H. M. and Fontaine, G. 1982 Astrophys. J. 254, 221.
Willson, L. A. and Bowen, G. H. 1984 Nature, 312, 429.

MULTIPLE POPULATIONS ON THE HORIZONTAL BRANCH

Robert T. Rood and Deborah A. Crocker

University of Virginia

1. INTRODUCTION

One of the more perplexing problems of HB morphology has been the discovery that the HB's of some globular clusters have a very long blueward extension separated from the main HB by a gap. This phenomenon was first noted by Cannon and Lee (1972) for the cluster NGC 6752. Similar gaps have also been found in the BHB of M 15 (Buonanno et al. 1984, who also provide a review of the topic) and NGC 288 (Buonanno et al. 1984). There is some indication of a BHB gap in M 2 (Harris 1975). A different bimodal distribution is found in NGC 2808 which has an HB populated on the red and blue ends but with a gap in the middle (Harris 1975, 1978). NGC 1851 (Stetson 1981), M 4 (Alcaino and Liller 1984) and NGC 6723 (Menzies 1974) also have HB's with relatively large populations of RHB and BHB as compared to the central HB. It is clear that for all of these clusters there are two "populations" of HB stars whether they are separated by a gap or not.

Buonanno, Corsi and Fusi Pecci (1984,85 and unpublished) have obtained high precision photometry of large samples of HB stars for a number of clusters. With data of this quality we can derive the mass distribution of HB stars. For example, with M 15 we find two more-or-less Gaussian distributions, centered at 0.67 $M_\odot$ and 0.59 $M_\odot$. The high mass group is much narrower ($\sigma \sim 0.01$ vs $\sigma \sim 0.025$). Once attuned to the possibility of multiple HB populations we find that even clusters without bimodal distributions in the CM diagram (e.g. M 3 and M 92) show some hint of having two superimposed mass distributions. Thus we will adopt the working hypothesis that that HB gaps are not gaps in the sense that some mechanism prevents stars from lying in that part of the CM diagram. Rather gaps occur when the two HB populations happen to be separated. Multiple populations may well exist in many clusters which show no gaps or bimodality.

Multiple HB populations could result if any of the HB "second parameters" varied within a given cluster. We reject age as a possibility because the required age difference (several billion years) seems far too large for stars within a given cluster. The other

A G Davis Philip, (ed)
Horizontal-Branch and UV-Bright Stars 99-105

possibilities include [CNO/Fe], Y, rotation or some other parameter affecting mass loss. For reference we refer to the redder HB population as group 1 (g1) and the bluer population as group 2 (g2).

2. [CNO/Fe]

Because HB color is primarily determined by CNO it is possible that the difference between g1 and g2 results from a difference in CNO. Because there is little spread in the RGB of M 15 at least in that case [CNO/Fe] must vary. Battistini et al. (1985) made the possibly crucial observation that the fraction of the HB made up of g2 stars seems to increase in the outer parts of M 15 (at about the 1 σ level.

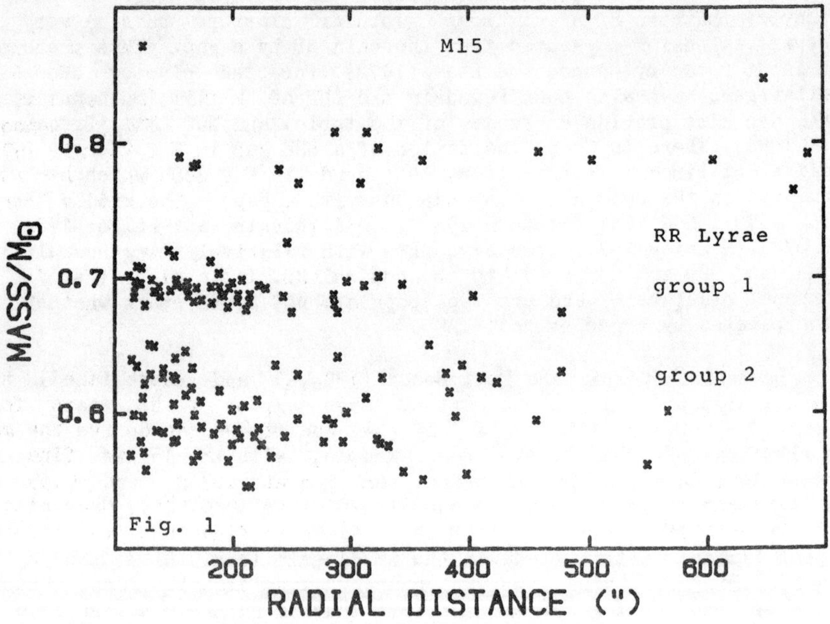

Fig. 1. The mass-radial position diagram for the HB stars of M 15 derived from the data of Battistini et al. (1985) assuming [Fe/H] = -2.0, [CNO/Fe] = 0 and Y = 0.25.

Figure 1 shows the masses derived from their data plotted as a function of distance from the cluster center. One scenario in which the HB morphology and radical distribution could be explained is the following. Suppose during the collapse of the protocluster some star formation begins. The stars in the outer cluster probably have not relaxed so their present orbits should reflect their initial orbits. At the same time a massive (perhaps supermassive) star forms at the cluster center. If it produces an O-rich supernova while the protocluster is still collapsing it could induce the formation of a second generation of O-rich stars which we now observe as g1. Thus g1 and g2 are the rough analogs of Populations I and II in the galaxy.

A simulated HB for such a cluster is shown in Figure 2. Here we have assumed that the mass loss efficiency is the same for the two groups. With [O/Fe] $\sim$ 0.6 for g1 and 0 for g2 two distinctly separate groups naturally arise. The g2 stars do not extend nearly as far to the blue as observed. To match the observations the mass loss efficiency for g2 stars must be 40% higher or the difference in [O/Fe] must be greater (Figure 3).

Such a picture has the virtue of producing a bimodal distribution naturally. In addition to a difference in spatial distribution of g1 and g2 stars it is likely that they would have different dynamical properties which might show up in a radial velocity survey. [CNO/Fe] also is the only parameter we have found which produces two distinct groups without an ad hoc increase in mass loss efficiency for g2.

3. HELIUM ABUNDANCE

A bimodal distribution in envelope Y of the HB stars could also produce the observed CM distribution. A bimodal primordial Y could result from a first generation of supermassive stars enriching the protocluster with Y-rich stellar winds. The amount of mass in such stars would have had to be substantial (several 10^4 $M_\odot$) to get the necessary ΔY. In this case the g2 stars would be the second cluster generation, the reverse of the [CNO/Fe] case.

A second possibility is that some parameter varies from star to star such that a dredge-up event occurs at the He-flash mixing He to the envelope in some stars. (It is imaginable that dredging could also mix out C, but in that case the dredged stars would become the redder g1 stars.) In either scenario the g2 stars must have a mass loss efficiency roughly 50% higher than the g1. Perhaps some additional mass loss is either associated with the dredge-up or with whatever parameter leads to dredge-up.

4. ROTATION

The most thoroughly discussed mechanism for producing a bimodal HB is rotation. A high rotation rate delays the He flash and allows extra mass loss to take place leading ultimately to bluer HB stars with higher core masses. Norris (1981) and Norris et al. (1981) find a bimodal distribution of CN strengths of the giants of M 4 and NGC 6752. They argue that the CN-strong stars are the result of meridional circulation mixing the products of CN burning to the surface in a group of rapidly rotating stars. These CN-strong giants should eventually evolve into the bluer g2 HB stars. Unfortunately Norris and Smith (1983) do not find a bimodal distribution of CN in NGC 2808. Also Peterson (1985)

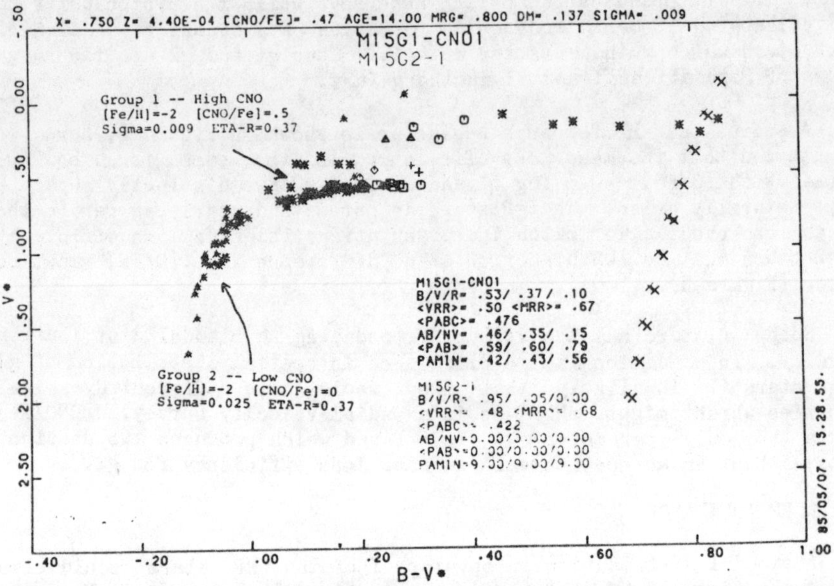

Fig. 2. A synthetic HB for M 15 with a bimodal distribution in [CNO/Fe].

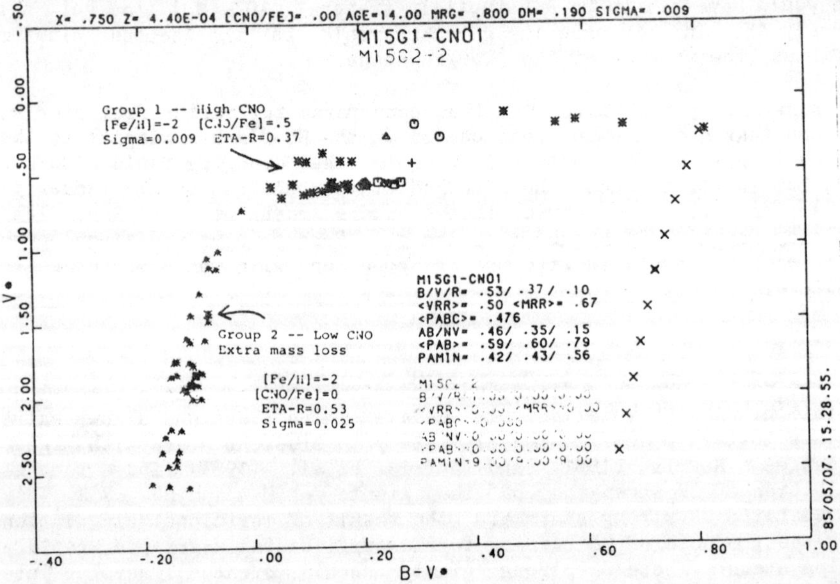

Fig. 3. Same as Fig. 2 except the mass loss efficiency for group 2 has been increased arbitrarily.

finds that stars from the supposedly more rapidly rotating g2 of M 4 are the most slowly rotating stars of any cluster that she has measured.

5. DISCUSSION

The origin of bimodal HB distributions remains unclear. The resolution of the problem will require observations in addition to color magnitude diagrams. The log g - log T_{eff} diagram may be particularly useful in distinguishing between the various possibilities. Figure 4 shows a CM diagram for a hypothetical cluster in which the g2 stars have arbitrarily increased mass loss. Such a situation might result form a distribution of magnetic field strengths if mass loss is predominantly via MHD wave-driven winds. Figure 5 shows a similar cluster in which the extra mass loss is associated with a rotation delayed He flash. Thus the core mass varies with total mass. There is little to distinguish the two mechanisms in the CM diagram. However, as seen in Figure 6, in the log g - log T_{eff} diagram not only is there a substantial shift of 0.07 in log g (at log T_{eff} = 4.2) for the rotation-mass loss case but a small slope change as well. For comparison the high helium g2 is shifted by about the same amount but the CNO produced g2 is shifted by only about 0.02 in the opposite sense.

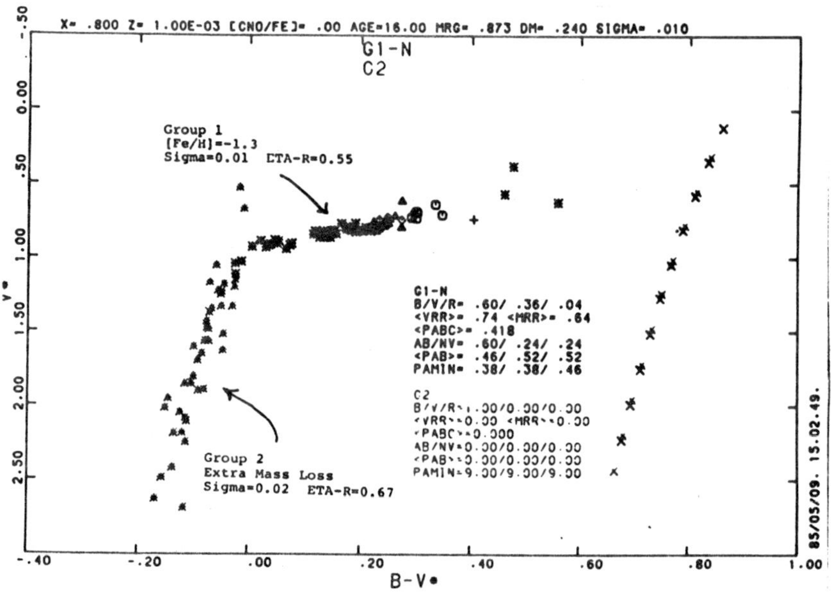

Fig. 4. A synthetic M 15 HB with a bimodal distribution of mass loss efficiency.

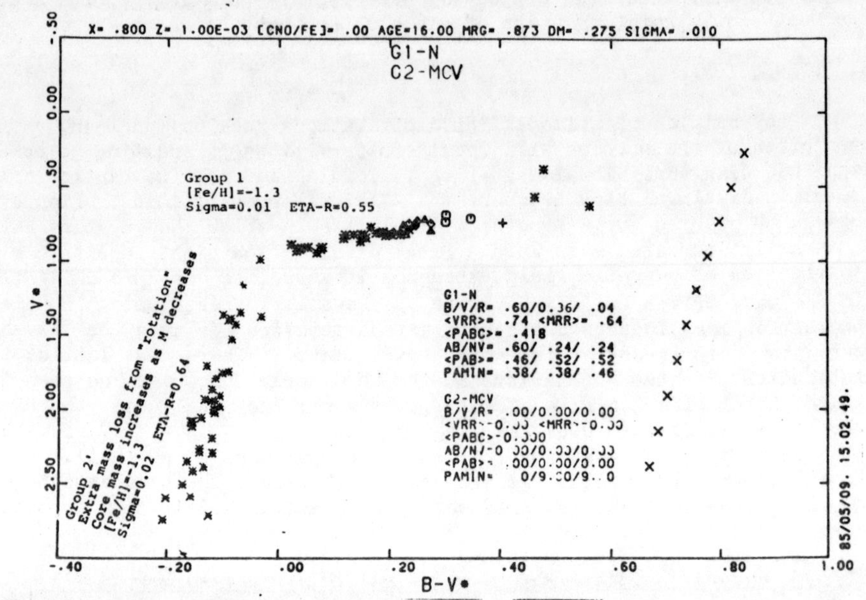

Fig. 5. Same as Fig. 4 except the high mass loss group is assumed to arise from rotation.

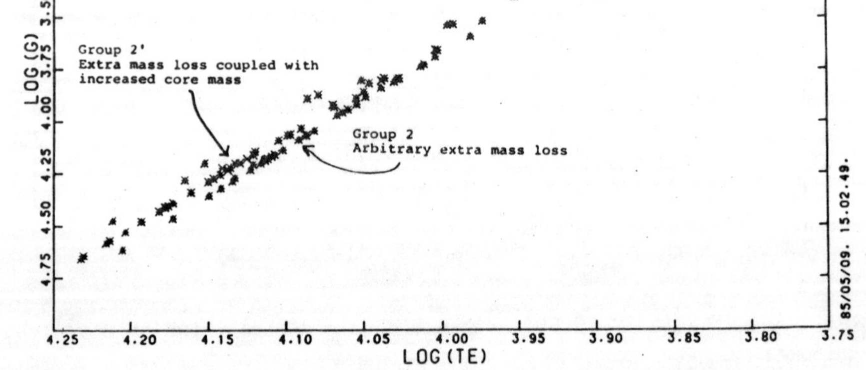

Fig. 6. The log g-log T_{eff} diagram for the stars shown in Figs. 4 and 5.

It is likely that several pieces of soft evidence will contribute to the solution. Direct measurements of abundances and rotation should be expanded. The tantalizing hint of different spatial distribution for g1 and g2 should be pursued and complemented by a radial-velocity survey. The effect of the different solutions on other stars in the CM diagram, while not as dramatic as on the HB, may be detectable. For example, the [CNO/Fe] mechanism predicts a color spread of ~ 0.02 - 0.03 in (B-V) at the turnoff tapering to less than 0.01 below.

We offer a final word of caution to observers of HB stars. Observing HB stars of one color virtually guarantees a biased sample. For example, low oxygen abundances in blue HB stars of a given cluster may give little evidence about oxygen in a typical cluster star.

REFERENCES

Alcaino, G. and Liller, W. 1984 Astrophys. J. Suppl. 56, 19.
Battistini, P., Bregoli, G., Fusi Pecci, F., Lolli, M. and Epps-Bingham, E. A. 1985 preprint.
Buonanno, R. Corsi, C. E. and Fusi Pecci, F. 1985 Astron. Astrophys. 145, 97.
Buonanno, R. Corsi, C. E., Fusi Pecci, F., Alcaino, G. and Liller, W. 1984 Astrophys. J. 277, 220.
Cannon, R. D. and Lee, S. W. 1972 unpublished (c.f. Cannon, R. D. 1984 in IAU Symposium No. 105, Observational Tests of the Stellar Evolution Theory, A. Maeder and A. Renzini, eds., Reidel, Dordrecht, p. 126.)
Harris, W. E. 1975 Astrophys. J. Suppl. 29, 397.
Harris, W. E. 1978 Publ. Astron. Soc. Pacific 90, 45.
Menzies, J. 1974 Monthly Notices Roy. Astron. Soc. 168, 177.
Norris, J. 1981 Astrophys. J. 248, 177.
Norris, J. Cottrell, P. L., Freeman, K. C. and Da Costa, G. S. 1981 Astrophys. J. 275, 120.
Peterson, R. C. 1985 Astrophys. J. 289, 320.
Stetson, P. B. 1981 Astron. J. 86, 687.

PUTTING THE "HORIZONTAL" BACK IN THE HORIZONTAL BRANCH

Deborah A. Crocker and Robert T. Rood

University of Virginia

ABSTRACT. We have devised a method for linearizing the observed horizontal branch (HB) of globular clusters. The theoretical zero-age HB (ZAHB), transformed to the observational plane, is used as a ridge line and the distance of a star along (X_{HB}) and the perpendicular (Y_{HB}) to that line is calculated. Having made this transformation the apparent mass distribution along the HB can be determined easily. The derived mass for a given star can be greater than the true ZAHB mass due to the evolution of the stars away from the ZAHB. However, this difference is usually very small. Therefore, the overall distribution of masses remains indicative of the true distribution. The distribution is also insensitive to small errors in E(B-V) and the distance modulus.

We have applied this technique to the data for the HB of M 15, principally to investigate the distribution around the gap in the blue HB. We find, using a standard model for M 15 (X = 0.75, Z = 0.0002, [CNO/Fe] = 0), that the mean mass of the blue group of stars is ~ 0.58 $M_{\odot}$ and the mass at the gap is ~ 0.65 $M_{\odot}$. The stars to the red of the gap have a significantly smaller mass dispersion than those to the blue. The same basic conclusion holds for [CNO/Fe] = 0.48; however, in this case the mass at the gap is inferred to be ~ 0.62 $M_{\odot}$. For the more metal-rich cluster, M 3, which has no gap, we find (using X = 0.75, Z = 0.0005, [CNO/Fe] = 0) that the mass distribution is much broader than either M 15 group and has a significant low mass tail. It is possible that M 3 has two mass populations, similar to those in M 15, but that they are not separated. By putting the HB into the X_{HB} - Y_{HB} plane it is also possible to get independent estimates of the distance modulus and reddening and to investigate details of HB evolution.

1. MOTIVATION

The standard representation of the horizontal branch in the V - (B-V) plane is unsatisfactory for several reasons. Some of the problems stem from the fact that, in this coordinate system the horizontal branch is not truly horizontal. For instance (B-V) may be used to derive a value for T_{eff} for stars near the RR Lyrae region but it is a very poor indicator of T_{eff} for the very blue HB stars where the HB is nearly

A. G. Davis Philip, (ed.)
Horizontal-Branch and UV-Bright Stars 107-116

vertical in the CMD. It is this tendency for the BHB to become vertical which makes it difficult to study the size and position of gaps which have been discovered recently in the BHB. In the V - (B-V) diagram it is also difficult to study the vertical structure of the HB as determined by the details of the evolution of the stars away from the zero age horizontal branch (ZAHB).

What is needed, then, is a measure which is a hybrid of both V and (B-V). The particular system we have used is the distance of a star along the ZAHB, (called X_{HB}), and the distance of a star perpendicular to the ZAHB (Y_{HB}). X_{HB} relates to T_{eff} (and mass) and Y_{HB} relates to the evolutionary progress of stars away from the ZAHB.

The masses derived from such a transformation are insensitive to small errors in the distance modulus and the reddening. On the other hand, large errors in DM and E(B-V) are quite apparent in results for X_{HB} and Y_{HB} and so, these can provide independent estimates of DM and E(B-V).

2. TECHNIQUE

We translate a HB star's position to the X_{HB} -Y_{HB} reference frame by using the ZAHB as the ridge line and projecting the star back on to it. The perpendicular distance is then Y_{HB} and the position along the ZAHB is X_{HB}. The zero point for X_{HB} has been arbitrarily set to where (B-V) = 0.

As we are using the ZAHB (as opposed to some empirical ridge line) we can use X_{HB} to derive quantities which are physically more meaningful (i.e. T_{eff} and mass). It is the value for mass which we derive which has proved to be most interesting. It should be noted, however, that the mass we derive is not the true mass but is distorted by evolution. As a star evolves it moves toward higher luminosity and becomes redder. Consequently it will be projected onto the ZAHB corresponding to a star of higher mass. The effect becomes more pronounced for redder stars. (In general, the stars which are shifted the most due to this effect can

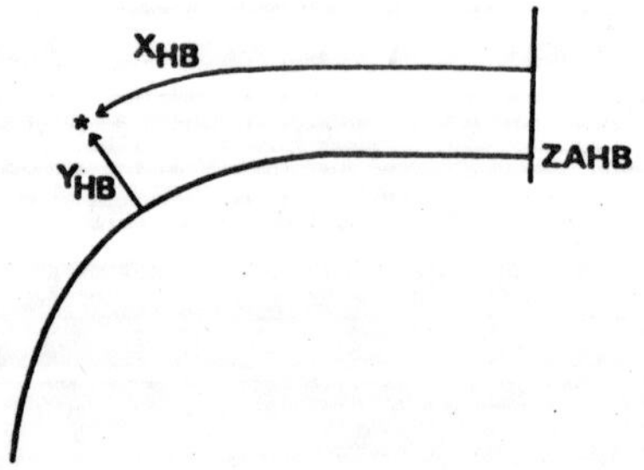

Fig. 1. Schematic of the Horizontal Branch

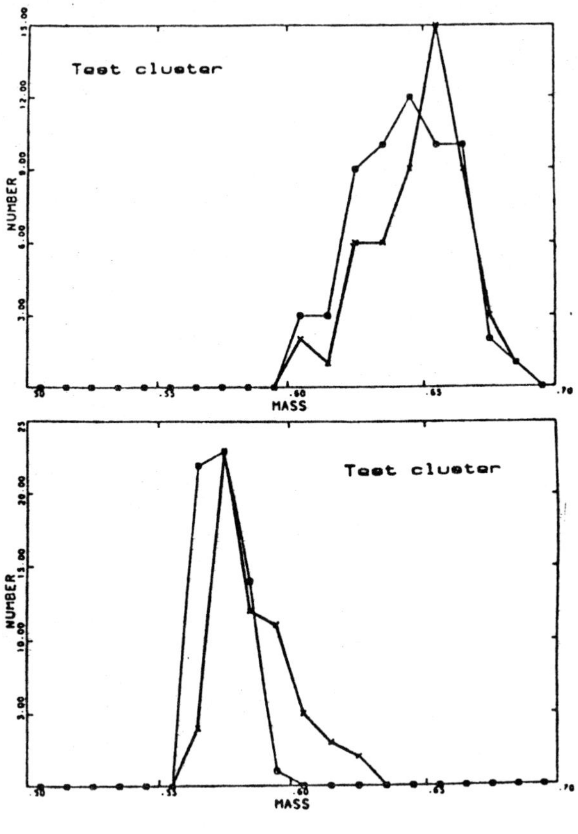

Fig. 2. Test Clusters, mass distribution. Top - mass loss 0.15 $M_\odot$
Bottom - mass loss 0.25 $M_\odot$

be identified because they are the ones with the largest values of Y_{HB}.

3. TEST CLUSTERS

We have produced a synthetic cluster (using our theoretical ZAHB and fitting HB evolutionary tracks), applied observation errors and then used this to rederive the mass based on X_{HB}. Illustrated in Figure 2 are the original mass distribution (x's) and the distribution arrived at through $X_{HB} - Y_{HB}$ (o's). The two panels represent the same composition (X = 0.75, Z = 0.0005) but different amounts of mass loss (top ∽ 0.15 $M_\odot$, bottom ∽ 0.25 $M_\odot$). The effects which were discussed previously can be seen in the test cluster.

4. THE REAL THING

We have applied the technique to the HB stars in M 3, M 5, M 15 and M 92. For M 3 we used a ZAHB with Z = 0.0005 and for M 5 we used Z = 0.001, for M 15 we used Z = 0.0002 and for M 92 a Z = 0.0001. In all cases, X = 0.75 was used. The data are from photometry which is statistically complete over a given area of the cluster. Therefore, our mass distributions are indicative of the true distribution (albeit with the bias due to evolution). At present, no RR Lyrae variables are included in the distributions.

4a. M 15

In this representation (Fig. 3) the gap in the BHB seems to be less a place where stars have migrated away (a true "gap") than the unpopulated area between two different distributions of stars. Here we show the results using the same ZAHB for the entire observed HB. One could imagine that the two groups did not share a common ZAHB. We have experimented with this but still find that the groups do not have the same mass distribution.

4b. M 92

The results for M 92, shown in Figs. 4 and 5, demonstrate a different power in using $X_{HB} - Y_{HB}$. The plot of $X_{HB} - Y_{HB}$ shows the results characteristic of a wrong value for DM. We find that DM ∽ 14.8 instead of the published value of 14.5 (Harris and Racine 1979). An alternative is that $Y < 0.25$ but the change produced by $\Delta DM = 0.3$ would require $\Delta Y \sim 0.1$ which seems unreasonable. In addition, the blue end of the HB drops below the ZAHB but this can most likely be attributed to a lack of photometric standards in that area (Buonanno et al., 1983).

4c. M 3

Fig. 6 shows the mass distribution for M 3. There is a significant low mass tail in the distribution. One can imagine easily that this result is like that for M 15 but in this case the two groups are not separated in mass space.

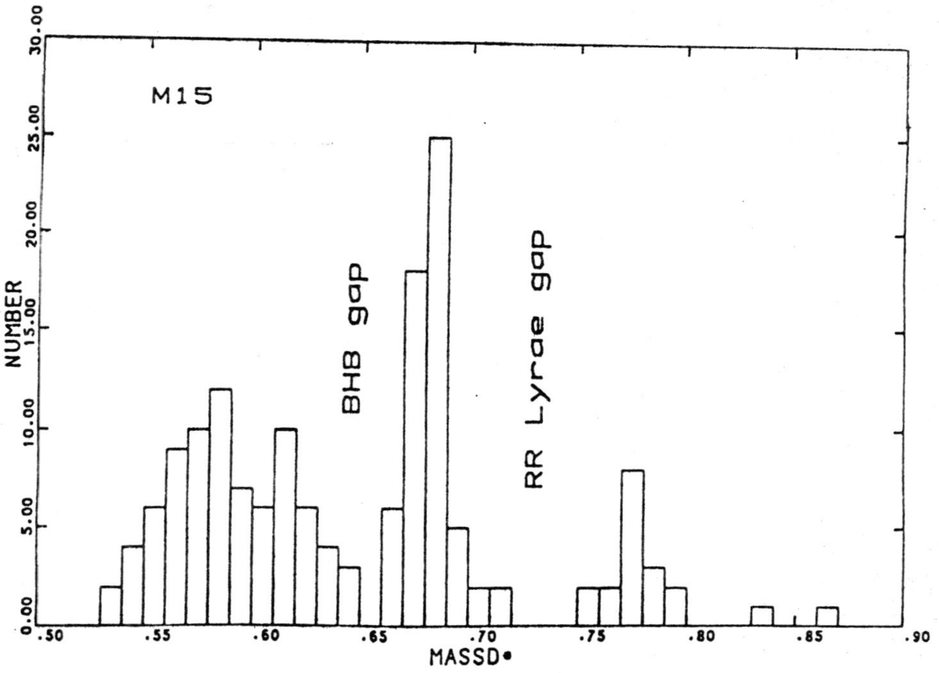

Fig. 3. The apparent mass distribution of M 15. The positions of the RR Lyrae gap and the BHB gap are indicated. The very blue (low mass) group of stars are not the tail of the high mass distribution but rather form a distinct group.

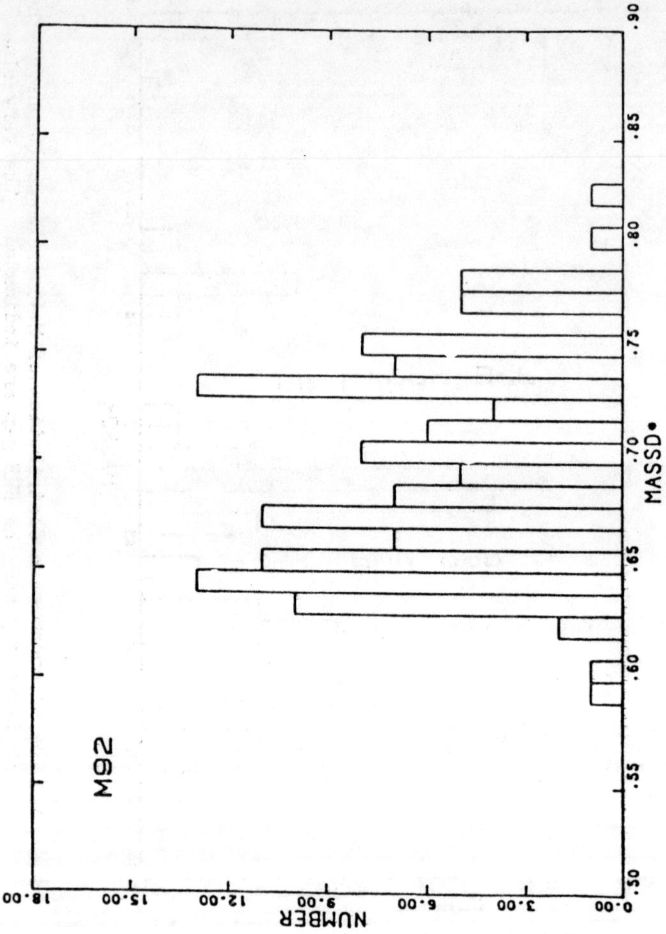

Fig. 4. The apparent mass distribution of M 92. The distribution may be bimodal.

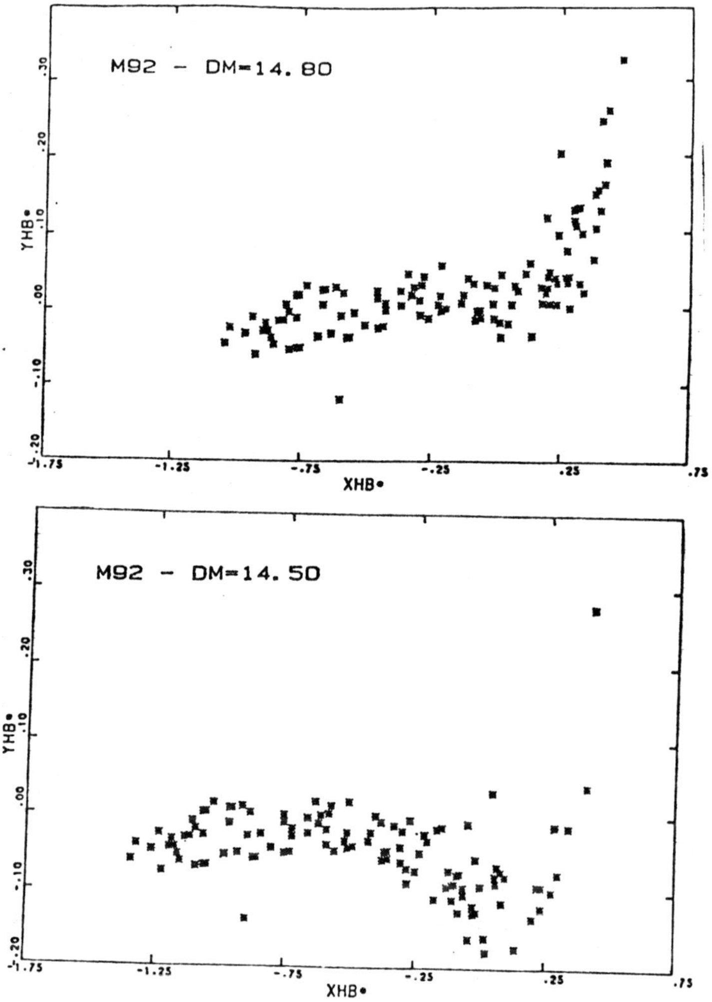

Fig. 5. The X_{HB} - Y_{HB} diagram for M 92. The lower panel shows the results using DM = 14.5, the currently published value. The top panel shows the result with the value revised to 14.8. The drop of the low mass end below the ZAHB is apparent in the upper panel.

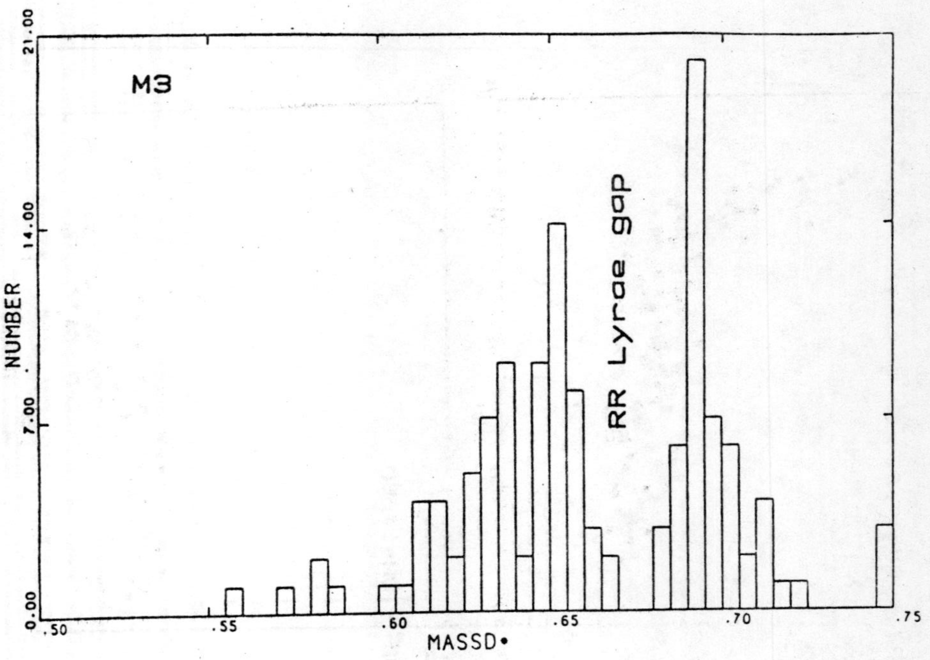

Fig. 6. The apparent mass distribution of M 3. There is a significant low mass tail on the distribution. The location of the RR Lyrae gap is indicated

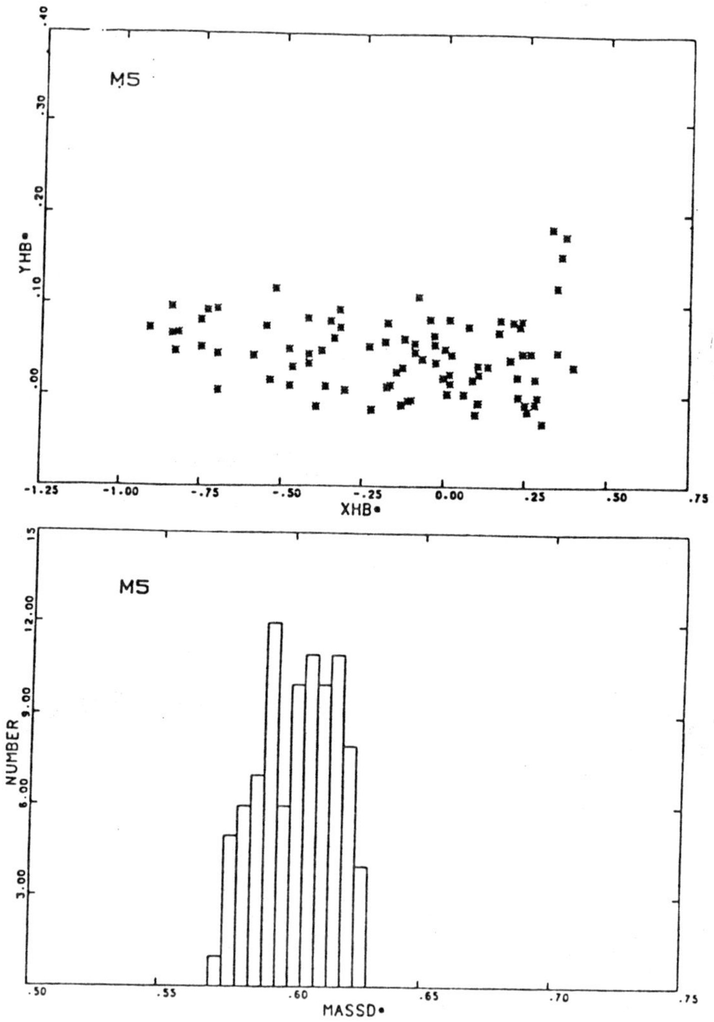

Fig. 7. The results of X_{HB} - Y_{HB} and the apparent mass distribution of M 5. The top illustration shows the tendency for the blue end of the HB to rise above the ZAHB. The lower diagram shows the mass distribution. Any stars to the red of the RR Lyrae gap are not included here.

4d. M 5

In Fig. 7 Y_{HB} versus X_{HB} and the mass distribution for M 5 are shown. There appears to be a slight rise in the blue end of the HB above the ZAHB. (This has been noted previously, c.f. Buonanno et al. 1981). Rotation has been suggested as the cause, however, this should also produce a low mass tail on the distribution. In fact, almost any mechanism working on the RGB, which increases Mc, would increase mass loss and produce a low mass tail. It may be however, that projection onto the wrong (non-rotational) ZAHB has caused a foreshortening of the distribution. The alternative explanation is that the standard colors are wrong for the very blue HB. In general, the results of the mass distributions for the four clusters seem to fall into three categories; two separated mass groups, as in M 15, two slightly merged groups, as in M 3 and M 92, and a single (or two fully merged?) group(s), as in M 5.

ACKNOWLEDGEMENT

We thank F. Fusi Pecci for providing data before publication.

REFERENCES

Battistini, P., Bregoli, G., Fusi Pecci, F., Lolli, M. and Epps Bingham, E. A. 1985 preprint. (M 15)

Buonanno, R., Buscema, G. Corsi, C. E., Iannicola, S. and Fusi Pecci, F. 1981 Astron. Astrophys. Suppl. 53, 1. (M 92)

Buonanno, R. Corsi, C. E. and Fusi Pecci, F. 1981 Monthly Notices Roy. Astron. Soc. 196, 433. (M 5)

Fusi Pecci, F. 1985 Private Communication. (M 3)

Harris, W. E. and Racine, R. 1979 Ann. Rev. Astron. Astrophys., 17, 241.

NOTE ON THE CHEMICAL ABUNDANCE OF EARLY-TYPE POPULATION II STARS

Keiichi Kodaira

Tokyo Astronomical Observatory

ABSTRACT. The present status of abundance analyses of early-type Pop II stars is reviewed briefly with an intention of clarifying the significance of HST observations of these stars.

1. EARLY-TYPE POPULATION II STARS

Observationally, two major groups of early-type Pop II stars can be identified; (1) Blue-Horizontal-Branch stars in the field and in globular clusters, (2) Hot subdwarfs, sdO and sdB stars in the field. The evolutionary status and the population properties of group (2) are not yet clear. A comparison of their locations in the luminosity-temperature diagram with theoretical loci suggests that a part of the sdB's can be on the extension of the BHB while sdO and other sdB's may be in the post-HB phase (cf. Drilling, Holberg and Schönberner 1984).

Newell (1973) added a minor third group, "high-luminosity Pop II B stars" which have luminosities several magnitudes above those of the BHB stars. They are apparently in the post-HB phase.

2. SIGNIFICANCE OF CHEMICAL ABUNDANCE ANALYSES

Knowledge of the chemical abundances of early-type Pop II stars is important for three reasons.

(1) The primordial and processed He abundance of Pop II stars can be inferred, preferably from early-type stars.

(2) Evolutionary effects may take place during the evolution from the giant branch to the HB or later. After mass-loss, one can observe the internal layers of giant stars as the surface of HB stars. Internal mixing of matter can affect the chemical composition derivable from spectroscopic observations.

A. G. Davis Philip, (ed.)
Horizontal-Branch and UV-Bright Stars 117-119

(3) Chemical peculiarities exist, similar to those in Bp-Ap and Am stars. The diffusion mechanism (cf. Michaud 1970) may be effective in the domain of early-type Pop II stars in the HR Diagram (cf. Baschek 1975). The comparison of chemical peculiarities between Pop I and Pop II stars in the same domain of the HR diagram may lead to an examination of the diffusion hypothesis and to a possibility of distinguishing the effects due to evolution from those due to diffusion.

At the present stage of analysis, these two effects are not yet separable. This situation seems to make it difficult to use the direct results of chemical abundance analyses of early-type Pop II stars as abundance indicators of population.

3. ABUNDANCE ANALYSES OF sdOB'S

A review of the work in the early period was given in Baschek (1975). Later major analyses appeared in Hunger et al. (1981) and in Heber et al. (1984), the latter utilizing both ground-based (ESO) and IUE observations. The analyses of spectra in the visible region and of low-dispersion spectra in the vacuum ultraviolet revealed abundance anomalies in He and C, which are depleted in sdB's.

High-dispersion UV spectra are very useful for fine abundance analyses of these hot stars (T_{eff} larger than 20,000 K). Baschek et al. (1982) analysed IUE high-dispersion spectra to show diverse peculiarities in the chemical abundance of sdOB stars. The depletion of He and Si in sdB stars may be due to diffusion effects while the depletion of C and O may be related to internal mixing.

4. ABUNDANCE ANALYSES OF BLUE-HORIZONTAL-BRANCH STARS

A review of the work in the early period appeared in Kodaira (1975). Recent studies by Kodaira and Philip (1984a,b) revealed the possibility that the diffusion mechanism, similar to that in Am stars, is operating in some hot BHB stars (T_{eff} about 10,000 K) in the field.

Pier (1983) also found a dozen Ca II K-weak stars among Curtis-Schmidt BHB candidates. Part of these stars were classified as Pop I Am stars. There seem to be additional Ca II K-weak stars with low metallicity among samples in his figure 7. These stars are similar to those found by Kodaira and Philip (1984b).

As yet no reliable analyses are available for He, C, N and O abundances of BHB stars. In order to obtain C, N, and O abundances of BHB stars one should analyse high-dispersion UV spectra.

5. ON HST OBSERVATIONS

In the case of sdOB stars, HST can improve both the spectral resolution and the depth reached in space, to disentangle the problem of evolution coupled with diffusion effects. The same is true also for the

BHB stars. HST can resolve hot early-type stars in the inner regions of globular clusters and may reveal any systematic trends in the abundance peculiarities with distance from the cluster center. Together with this unique role for HST in this research area, much more high-resolution work is necessary to draw full advantage from HST observations.

REFERENCES

Baschek, B. 1975 in Problems in Stellar Atmospheres and Envelopes, B. Baschek, W. H Kegel and G. Traving, eds., Springer, Heidelberg, p. 101.

Baschek, B., Hoflich, P. and Scholz, M. 1982 Astron. Astrophys. 112, 76.

Drilling, J. S., Holberg, J. B. and Schönberner, D. 1984 in Future of Ultraviolet Astronomy Based on Six Years of IUE Research, NASA CP-2349, p. 249.

Heber, U., Hunger, K., Jonas, G. and Kudritzki, R. P. 1984 Astron. Astrophys. 130, 119.

Hunger, K., Gruschinske, J. Kudritzki, R. P. and Simon, K. P. 1981 Astron. Astrophys. 95, 244.

Kodaira, K. and Philip, A. G. D. 1984a Astrophys. J. 278, 201.

Kodaira, K. and Philip, A. G. D. 1984b Astrophys. J. 278, 208.

Lamontagne, R. M., Wesamael, F., Fontaine, G. and Sion, E. M. 1984 in Future of Ultraviolet Astronomy Based on Six Years of IUE Research, NASA CP-2349, p. 254.

Michaud, G. 1970 Astrophys. J. 160, 641.

Newell, E. B. 1973 Astrophys. J. Suppl. 26, 37.

Pier, J. R. 1983, Astrophys. J. Suppl. 53, 791.

ULTRAVIOLET OBSERVATIONS OF POPULATION II OBJECTS WITH THE HST: CLUSTERS OF STARS (GLOBULAR CLUSTERS AND ELLIPTICAL GALAXIES)

Klaas S. de Boer *

Kapteyn Astronomical Institute

1. INTRODUCTION

The launch of the Hubble Space Telescope (HST) will provide the astronomical community with a first class telescope in a perfect observational climate. The very fine optics, with the capability of observing objects in the entire wavelength range from the visible infrared to the region of the Lyα resonance line, will open a multitude of possibilities for in-depth investigations of a large variety of objects. The purpose of the papers collected here is to point at observational projects where the HST may contribute essential information to answer questions in the research of Population II objects.

The Population II objects we have interest in are horizontal-branch stars in the field of the galaxy (FHB) and those in globular clusters (the blue ones being more conspicuous (BHB) than the red ones), the luminous but exceptional post-AGB stars, and conglomerates such as globular clusters and elliptical galaxies which will show a blend of different kinds of Population II stars. The understanding of the stars is an essential prerequisite for understanding the clusters and galaxies, including the step to the color change of redshift blended with intrinsic evolution. We wish to know about metal abundance and how it may affect the spectral energy distribution, about the He abundance and its relation with cosmology. We would like to investigate mass loss and the rotation of such stars. A certain amount of information is available from the visual. Actually, more spectroscopic detail, together with better atomic data and spectral line analysis would improve our understanding of such stars substantially.

We must ask ourselves where (in which wavelength range) will we gain the most information in the coming years. And where is the HST

---

* On Assignment at Royal Greenwich Observatory

A. G. Davis Philip, (ed.)
Horisontal-Branch and UV-Bright Stars 121-127

essential to further our understanding? Relatively large amounts of time could be used on Earth-based telescopes to increase our data bases with hitherto unknown detail. Calculations of radiation transfer through atmospheres of metal-poor evolved stars may bring discoveries and bring better understanding of spectra and spectral energy distributions. For the probing of the ultraviolet part of the spectrum soon the HST will be available as an observatory. Note, however, that the UV has a disadvantage for absolute spectrophotometry because of the UV extinction and its non-uniform spectral behavior in the galaxy. New information concerning FHB and BHB stars would come from the use of the HST in its various modes. This excludes studies in the infrared, where the HST would not have real advantages over ground-based observations, and possibly some kinds of studies in the visual which can be done from ground-based observatories as well. The ultraviolet part of the spectrum is most useful for the hotter Population II objects such as BHB and PAGB stars. Stars of temperatures near to 10,000 K will have an extremely rich UV spectrum, as can be seen in the 200-300 nm range spectrum of HD 109995 obtained with the International Ultraviolet Explorer (IUE) high resolution spectrograph. For stars at quite higher temperatures, the number of spectral lines available for observation may be smaller. The high ionization stages of elements have the majority of the resonance transitions shortward of the Lyman continuum, and only the weaker high excitation transitions may be available. These are weak (see de Boer 1985) because of the low metallicity of Population II stars.

Calculations of model atmospheres for metal-poor stars are available from Kurucz (1979 and later work). A comparison with spectral energy distributions (IUE) of FHB stars showed that in the UV the opacity tables need adjustment (Huenemoerder, de Boer and Code 1984). The models predicted too little flux in the 250 nm range and too much shortward of 150 nm. Also for some PAGB stars a UV spectral energy distribution was observed for which no matching model could be found (de Boer 1985).

Let us address some object types and estimate what one can learn.

2. STARS IN THE FIELD

FHB stars have to be found and identified in the field of the galaxy. From visual colors one can identify the cool FHB stars ($T <$ 9000 K) while very blue objects may be sdB stars (but see also Tobin 1984, 1985, who shows that many high latitude blue stars are of Population I). The intermediate group must be identified from spectra. There is a restriction on the worth of the parameters derived for these stars; since they are evolved stars their surface metallicity may be modified from their initial composition, while their ages cannot be determined individually. In the UV the spectral energy distribution is different from that of Population I stars (de Boer and Wesselius 1981;

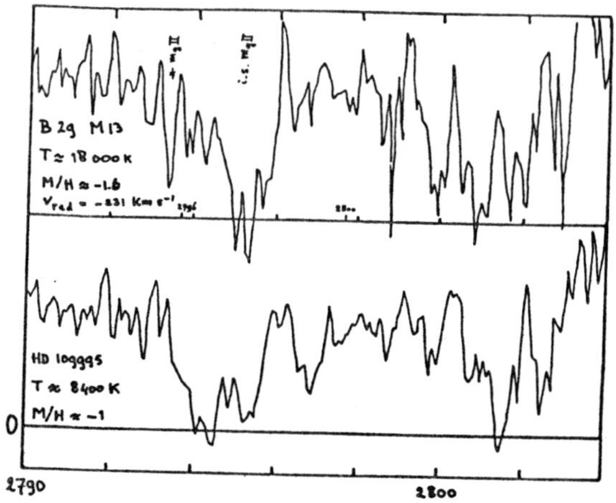

Fig 1. The 280 nm region of the spectrum of B29 in M13 (de Boer and Savage 1983) and of the FHB star HD 109995 (de Boer, unpublished). Both IUE spectra are rather noisy; B29 because of V=13.0, HD 109995 because of enhanced image-tube noise during the observation. Low resolution spectra of FHB and other population II stars can be found in Huenenoerder et al. (1984) and Cacciari (1985), respectively.

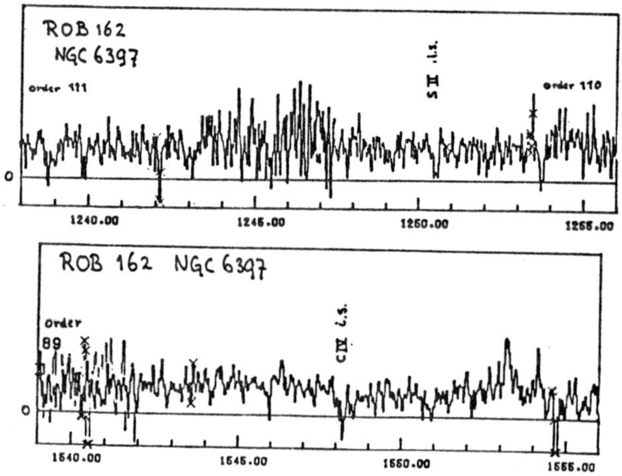

Fig 2. Two portions of the spectrum of ROB162 in NGC6397 (de Boer, unpublished). Apart from noise (V=13.2 E(B-V)=0.2) and interstellar lines the spectrum seems featureless.

Huenemoerder, de Boer and Code 1984). A portion of an IUE high dispersion spectrum of HD 109995 is shown in Fig. 1.

3. STARS IN GLOBULAR CLUSTERS

For star clusters both metallicity and age are approximately known from current ground-based studies. This sets the framework for observations of cluster stars.

3.1 Red Giants

The peak of their radiation lies in the infrared so most studies have to be carried out from the ground, at least as long as the HST has no IR facility.

3.2 BHB Stars

With the original metallicity known, the spectral lines will tell us about chemical processing, convection, dredge-up, etc., in the evolutionary phases between red giant and white dwarf. Spectra of these stars have been observed from the ground (Kodaira and Philip 1984) but the HST has some advantage in the quality (but maybe not quantity) of the data. In particular the hotter BHB stars are best observed with the HST because of their larger intrinsic brightness there (watch out for extinction!).

3.3 Main Sequence Stars

These stars will be best comparable with models since we think we understand the structure of stars on the main sequence. They will provide good tests for calculations of model atmospheres with substantially reduced metallicity. However, in globular clusters these stars are solar like and thus so faint (at cluster distances) that the HST also will have difficulty in getting useful data.

3.4 PAGB Stars

The enhanced luminosity of these above BHB stars makes them easy targets for observation. They form the tie to stars which are the nuclei of planetary nebulae. Again, they have an evolved atmospheric composition (see e.g. PN in M 15, Auriere et al. 1984). Low resolution IUE spectra exist for several of them (de Boer 1985) and three have been observed with IUE at high dispersion (B 29 in M 13, vZ 1128 in M 3, ROB 162 in NGC 6397; see de Boer 1985, and Figs. 1 - 4).

4. POPULATIONS IN GLOBULAR CLUSTERS

Images of globular clusters at a set of ultraviolet wavelengths will provide information on broadband colors of all stars in a cluster. Such studies may show how extensive the population on the BHB is (Bohlin et al. 1985) and will help to locate the post-PAGB / PN-nuclei / pre-WD

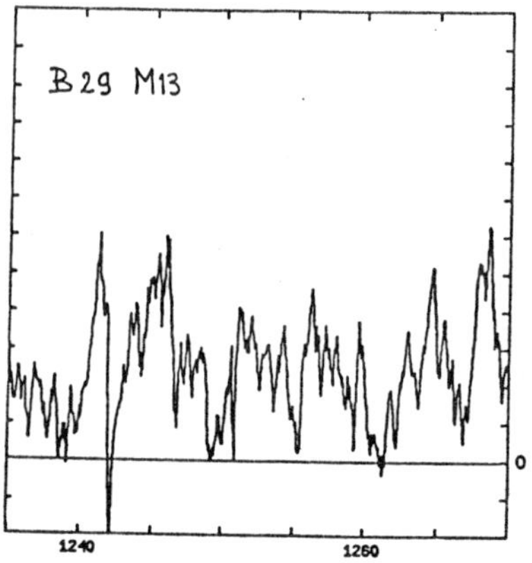

Fig 3. A portion of a smoothed IUE spectrum of B29 in M13 (de Boer and Meade, unpublished).

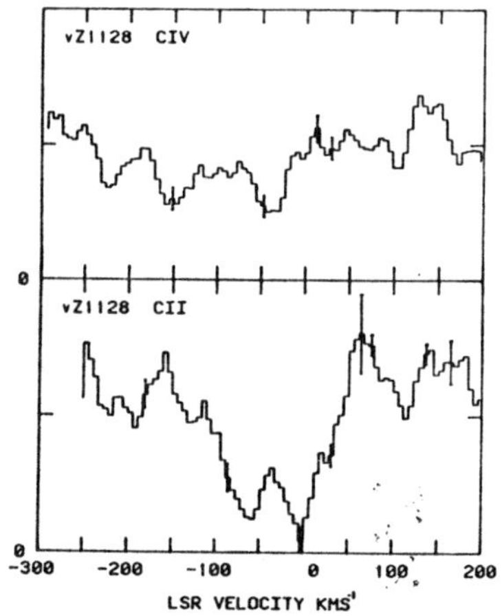

Fig 4. The CIV and CII region of the spectrum of vZ1128 in M3 (de Boer and Savage 1984) observed for interstellar lines. Note the CIV absorption at cluster radial velocity.

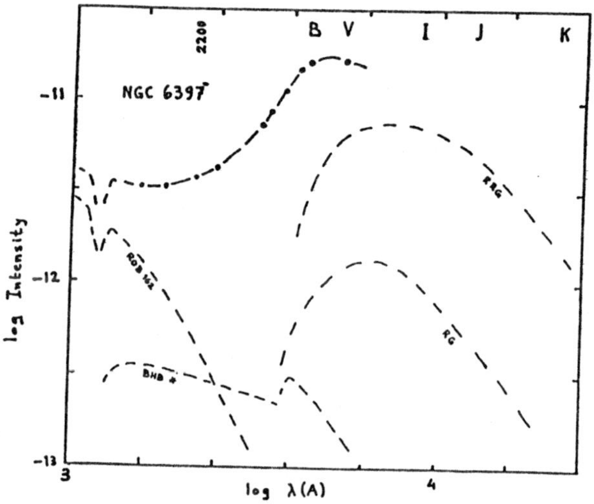

Fig 5. Population synthesis for NGC6397 based on integrated infrared, visual and UV colours. Input data are from sources as given by de Boer (1985), who also presented a similar diagram for M13.

stars.

When the spectral energy distribution of all constituents of a globular cluster is known, one can attempt to synthesize spectra. In the visual this has been done mostly based on photometry, which may include some line indices (Burstein et al. 1984). In the UV the synthesis has relied on broad-band colors (Welch and Code 1980; de Boer 1985, and see Fig. 5). Only with synthesized energy distributions can one hope to understand much larger and less well-defined systems such as the nuclei of galaxies (Welch 1982) and elliptical galaxies.

5. CONCLUSIONS

Population studies are essential to the understanding of integrated light from galactic globular clusters (van Albada, de Boer and Dickens 1981), Magellanic Cloud globular clusters (de Boer 1981, Cohen, Rich and Persson 1984) and elliptical galaxies (de Boer 1982). It is therefore of utmost importance to observe individual cluster stars and get absolute spectral energy distributions from the visual to the farthest available UV wavelengths. In addition, accurate photometry should be carried out for a sample of stars in globular clusters.

Population syntheses can be carried out with such visual-ultraviolet data bases. One can then calculate the colors of elliptical galaxies at various stages in their evolution. This will provide insight in the color versus redshift behavior when probing the distant reaches of the universe.

REFERENCES

Auriere, M, Adams, S and Seaton, M. J. 1983 Monthly Notices Roy. Astron. Soc. 205, 571.
Bohlin, R. C., Cornett, R. H., Hill, J. K., Smith, A. M, and Stecher, T. P. 1985 Astrophys. J. 292, 687.
Burstein, D, Faber, S. M., Gaskell, C. M. and Krumm, N. 1984 Astrophys. J. 287, 586.
Cacciari, C. 1985 Astron. Astrophys. Suppl. In press.
Cohen, J. G., Rich, R. M. and Peterson, S. E. 1984 Astrophys. J. 285, 595.
de Boer, K. S. in IAU Colloquium No. 68, Astrophysical Parameters for Globular Clusters, A. G. D. Philip and D. S. Hayes, eds., L. Davis Press, Schenectady, p. 3.
de Boer, K. S. 1982 Astron Astrophys Suppl. 50, 247.
de Boer, K. S. 1985 Astron. Astrophys. 142, 321.
de Boer, K. S. and Savage, B. D. 1983 Astrophys. J. 265, 210.
de Boer, K. S. and Savage, B. D. 1984 Astron Astrophys. 136, L7.
de Boer, K. S. and Wesselius, P. R. 1981 Astron. J. 85, 1356.
Huenemoerder, D. P., de Boer, K. S. and Code, A. D. 1984 Astron. J. 89, 851.

Kodaira, K. and Philip, A. G. D. 1984 Astrophys. J. 278, 201.
Kurucz, R. L. 1979 Astrophys. J. Suppl. 40, 1.
Tobin, W. 1984 Astron. Astrophys. Suppl. 57, 427.
Tobin, W. 1985 Astron. Astrophys., in press.
van Albada, T. S., de Boer, K. S. and Dickens, R. J. 1981 Monthly Notices Roy.l Astron. Soc. 195, 591.
Welch, G. A. 1982 Astrophys. J. 259, 77.
Welch, G. A. and Code, A. D. 1980 Astrophys. J. 236, 798.

PARALLAXES AND THE OLD-POPULATION DISTANCE SCALE

Arthur R. Upgren

Van Vleck Observatory

The distance scale for old-population objects is rather independent of that for young objects but it ultimately rests upon the parallaxes of nearby representative stars. A study by van den Bergh (1977) summarizes the degree of independence of the young and old star distance scales. In recent years, the distance scale for young stars has improved greatly through the parallax observations of the Sproul, Van Vleck and U. S. Naval Observatories and others for field stars and for the distance calibration of the Hyades and other clusters. But the improvement of the old-star distance scale has not kept pace. This is because there are no nearby old clusters and because astrometric programs have neglected the relatively few old subluminous field stars which are close enough for good parallax determinations to be made. It also suffers from persisting uncertainties in analyses of statistical parallaxes of field RR Lyrae stars. This is perhaps best illustrated by Sandage (1970) who derived the distances to four globular clusters (M 3, M 13, M 15 and M 92) from a comparison of their main sequences with that defined by nearby field stars with high ultraviolet excesses. For the calibration of the extreme uv-excess objects, he was limited to only nine stars with trigonometric parallaxes which were not of high precision. Later, Hanson (1979) reanalyzed these same parallaxes and found a different locus for the main sequence which they defined. In both cases, the paucity of the data allows considerable uncertainty in the result.

The Van Vleck Observatory has as one of its astrometric goals the determination of precise parallaxes of subluminous old stars in order to improve their distance scale. About 50 stars with moderate to extreme ultraviolet excesses have been observed for parallax for several years, and for eight of these, observations and measures are completed and published. The others will be completed over the next two or three years.

Existing available parallaxes of old stars from all sources show a great increase in total parallax weight over the last few years. Other parallax efforts on old stars are expected later from both the Space Telescope and the Hipparcos Satellite of the European Space Agency. It is not yet clear whether and by how much either satellite will improve

A. G. Davis Philip, (ed.)
Horizontal-Branch and UV-Bright Stars 129-131

the old-star distance scale, but comparison with the present ground-based data will be useful in any event in the calibration of the zero-point of the total system of parallaxes.

Since the kinematics of the presumed old stars are the data being sought, the non-kinematical parameters are best used to illustrate the calibration problem and its improvement. The F, G and early K dwarfs are the stars on which the old-star main sequence is based and to which the globular cluster main sequences can be compared. Sandage (1969) has developed the normalized ultraviolet-excess parameter, $\delta\ 0.6(U-B)$ which is asymptotic, reaching a maximum value of about 0.31 and applicable for stars with $0.35 < (B-V) < 0.80$ (F2V to K0V on the MK system). For most of these middle-range stars, $\delta\ 0.6(U-B)$ differs only slightly from the simpler excess parameter $\delta\ (U-B)$. Carney (1979) has shown a close relation between these indices and [Fe/H], a more physically significant quantity, defined as log (Fe/H; star) - log(Fe/H; Sun).

The calibration by Sandage (1970) was based on only 9 stars with $\delta\ (0.6) > 0.16$ and another 16 stars with $0.11 < \delta\ (0.6) < 0.15$, which are in his excess groups C and D, respectively. Nearby stars with smaller excesses than 0.11 and with accurate trigonometric parallaxes larger than 0".040 are abundant throughout the F, G and K main sequence (Gliese, Jahreiss and Upgren 1985) but parallaxes for stars with high excesses are rare. From Carney's (1979) relation, Sandage's C and D groups correspond to [Fe/H] values from about -0.4 to -0.7, and more than -0.7, respectively. The stars of the solar neighborhood are distributed in [Fe/H] somewhat normally with a mean near 0.0 and a dispersion of about ±0.2 dex (e.g. Twarog 1981) so we can expect very few nearby stars with [Fe/H] < -0.5. The lower limit in metallicity of the stars of the disk population must be near this limit of -0.4 to -0.5.

Sandage found that the stars in his C and D groups average 0.6 and 1.2 magnitudes, respectively, below the main sequence formed by the Hyades and similar metal-rich stars. Hanson's (1979) reanalysis produced a somewhat smaller deviation from the metal-rich main sequence. The difference was due in part to Hanson's introduction of necessary systematic corrections to the observed trigonometric parallaxes and partly due to the insufficient data. The most significant consequence of the uncertainty was that the distances of the globular clusters are increased by 0.4 mag. or 45 percent and that the mean absolute magnitudes of RR Lyrae stars in those clusters move close to zero.

Hanson's rationale is sound but the data available to him and to Sandage were so poor that the principal necessary systematic correction to parallaxes, the Lutz-Kelker correction, could not be applied to most of even those few stars. This correction is limited to very good parallax data ($\sigma(\pi)/\pi < 1/6$ which leads to $\sigma(M) < 0.3$ mag.) and Hanson was forced to deal with corrections to the proper motions of most of the stars.

The continuing uncertainty in the distance scale for old objects has prompted the Van Vleck Observatory to include a large number of nearby subdwarfs and subdwarf candidates among the stars on its parallax program. These include a list of 18 suspected subdwarfs in 16 systems identified by Strand (1958) from their published parallaxes and proper motions and proposed by him as standard stars for all parallax observers. This list of stars, with observations made for parallax by many participating observatories, was only recently published (Upgren 1985). But the wealth of available astrometric data for them and their confirmation as subdwarfs (Carney 1981) promises parallaxes of high precision. They qualify for Sandage's D group and some of them were among the first subdwarfs to be published in the course of the astrometric program at Van Vleck. They appear in its most recent list of parallaxes (Upgren et al. 1985). This list is also the first in its series to incorporate measures made on the PDS microdensitometer at Yale University. The expected external parallax error with the PDS machine (not yet fully confirmed) is about ± 0.006 as opposed to ± 0.008 for earlier Van Vleck parallaxes measured by hand, and ± 0.016 for most of the stars available to Sandage for his calibration.

The greater number and higher precision of the subdwarf parallaxes now available can be used to evaluate the Lutz-Kelker correction and lead to a more trustworthy old-star main sequence. Sandage had only two stars in each of his two groups (C and D) with the high level of precision required for the correction, whereas twelve such stars are now available and as many more will become available in a year or two (assuming each new subdwarf parallax turns out to be close to the photometric parallax predicted by the V magnitude and (B-V) color). Many other subdwarfs will have parallaxes which are less precise but still useful in the calibration. The distances to globular clusters may be changed or not, but in either event, they will soon rest on a much firmer foundation.

REFERENCES

Carney, B. W. 1979 _Astrophys. J._ 233, 211.
Carney, B. W. 1981 Private Communication.
Gliese, W., Jahreiss, H. and Upgren, A. R. 1985 in _The Galaxy and the Solar System_, R. Smoluchowski, J. N. Bahcall and M. S. Matthews, eds., University of Arizona Press, Tucson.
Hanson, R. B. 1979 _Monthly Notices Roy. Astron. Soc._ 186, 875.
Sandage, A. R. 1969 _Astrophys. J._ 158, 1115.
Sandage, A. R. 1970 _Astrophys. J._ 162, 841.
Strand, K. Aa. 1958 _Astron. J._ 63, 152.
Twarog, B. A. 1981 _Astron. J._ 86, 386.
Upgren, A. R. 1985 _Observatory_ 105, in press.
Upgren, A. R., Weis, E. W., Nations, H. L. and Lee, J. T. 1985 _Astron. J._ 90, 652.
van den Bergh, S. 1977 _Astrophys. J. Letters_ 215, 103.

GENERAL DISCUSSION

RENZINI: For most of the clusters, they are larger than the view of the wide field camera.

PHILIP: I have heard that there are no Strömgren Filter sets on Space Telescope.

LECKRONE: The FOC, faint object camera, does. The high speed photometer does not.

PHILIP: Saul, can we get enough resolution for abundances at 25,000?

ADELMAN. Yes

LECKRONE: At 25,000 one would be working in the first order and there is very little scattered light, very little background. The signal to noise level should be quite superior. 25,000 is about a factor of two better than IUE and a considerably better signal to noise ratio than IUE. So the quantitative analyses will be much better.

ADELMAN: But, if you have any rotation whatsoever it does not necessarily make sense to go to 95,000 resolution.

LECKRONE: Under 10 km/sec it is still possible to use the 95,000 resolution.

PHILIP: Saul, you were telling me that [Fe/H] for HD 109995 was going down?

ADELMAN: Yes, relative to the original Kodaira, Greenstein and Oke value. Determining its gravity is difficult becuase the H gamma profile is temperature sensitive. We need to find a gravity sensitive indicator, perhaps in the energy distribution.

PHILIP: Up to the present time, if you look at the FHB stars for which spectroscopic abundances have been done, they all fall near [Fe/H] = -1 to - 1.4 or so. What is odd is that if you look at the globular cluster abundances you go from [Fe/H] = -1 to -2 so the question is, "Where are the [Fe/H] = -2 FHB stars.". When Saul told me that the abundance of HD 109995 was going down, this was great for we would then have one very metal-poor FHB star.

LECKRONE: Saul, do you have a problem with Feige 86?

ADELMAN: The photospheric abundance of F 86 should have been modified by diffusion.

JANES: If you can answer the problem with any other instrument then you can't do it with space telescope.

ADELMAN: In principle, one can try to get CNO abundances from infrared lines with CCD detectors such as those at KPNO. It is possible to get the oxygen abundance from the ground for the brighter stars but the nitrogen lines are weaker and the carbon lines are contaminated by lots of telluric lines. Another nasty problem is NLTE which differs from line to line.

LECKRONE: There are much longer excitation lines. Surely we are going to get some oxygen abundances from these infrared lines but the question is, "How accurate are they?". The only way to find out is go looking at these and see if they agree. Maybe just a few cases would be sufficient.

JANES: So one could do abundances with this set of lines without worrying about Non LTE?

LECKRONE: Yes. Whatever NLTE effects there are ought to show up because you cover a pretty wide range of excitation between the optical and the UV. So you are looking for self consistency. Secondly, because they are high excitation lines with high populations you can have a 1% NLTE effect in these lines which you could not even see but that could translate into a 50% NLTE effect in the higher excitations.

JANES: We could use the optical high excitation lines to see how badly we were off in NLTE.

ADELMAN: For some O I lines calculations have been performed by Baschek, Scholz and Sedlmeyer (1977 Astron. Astrophys. 55, 328.) but how well they apply to Pop II is a question.

RENZINI: Any kind of abundance determination is interesting because primordial nucleosynthesis in the galaxy is important to study.

LECKRONE: Using the FHB stars as a starting point one could try to determine the metallicity of the fainter BHB stars.

PHILIP: When you pick a bunch of globular clusters you already know the [Fe/H] range.

JANES: How likely is it that FOS spectrophotometry sections can be matched so as to obtain good absolute spectrophotometry ?

LECKRONE: There is a very extensive calibration program, not on the ground, but in orbit which is tied into the IUE scale and the idea is to use astronomical standards to calibrate the spectrograph. In the optical region there is a set of standards that Oke is determining.

JANES: The reason that I am asking that is that the range of plausible masses is rather small. If you are able to get really good absolute spectrophotometry you could get temperatures and gravities and therefore get another hold on the distance scale.

RENZINI: However, you do not have so many stars that are so blue.

LECKRONE: This is out of Huenemoerder et al. and it shows FHB stars relative to Pop I early-type stars. In just that part of the far uv wavelength the flux distribution is extremely sensitive to metallicity and if you can tie together the flux distribution from the UV to the infrared it seems to me it gives you a very strong leverage on determining simultaneously effective temperature, gravity, metallicity, bolometric luminosity and so forth. If you can not fit that entire range of data then there is a big problem that has to be solved.

JANES: The question is if one can determine all these parameters accurately enough to add something new.

LECKRONE: If you have done it for the FHB stars, of what interest is it to push out to the globular clusters and try to do the same thing for BHB stars. Or should we accept Philip's hypothesis that the FHB and BHB stars have such a good match that once you solve the problems for the FHB stars you don't need to go to the BHB stars.

JANES: You would like to solve it for a particular globular cluster.

PHILIP: Right. We do have globular clusters that do not match the FHB metal abundance distribution. Until this metal abundance problem is solved we don't have any metal-poor FHB stars so you would have to go to the globular clusters in order to study the low-metallicity group.

JANES: Why are not there any very metal-poor FHB stars?

PHILIP: In the 'observed bag' so far they do not seem to exist. Whether that is a real effect or just a selection effect in that we have picked out only the more metal-rich FHB stars is not known at the present time.

RENZINI: Can the metallicity be determined by the m_1 index?

PHILIP: No. For A0 to about A3 the change in m_1 with [Fe/H] is so small that it is less than the measuring error. Only for stars of spectral type A5 and later can the delta m_1 index be used to determine metallicity. For early-type A stars the only way to determine metal abundance is to do a spectroscopic line analysis.

INDICES

If a page number is underlined it indicates that the reference is to an author of a paper (Name Index) or to a subject listed in the title of a paper (Object and Subject Indices). Objects are listed in the Subject index if the object is mentioned in the title of the paper or in a chapter heading. Other mentions of objects are indicated in the Object Index.

NAME INDEX

OBJECT INDEX

SUBJECT INDEX